Great Britain

The mother of invention

Its contribution to the
modern world

This edition first published
in June 2018 by

Greentech publishing

Keith Chamberlain

Printed and bound in the UK

ISBN: 978-1-9995948-0-0

A CIP catalogue record for this book is available from the British library.

This book can be ordered direct from the publishers at: sales@greentechpublishing.com

www.greentechpublishing.com

Acknowledgements

I would like to thank the people, too numerous to mention, who have encouraged and mentored me over the years. To my parents, teachers and lecturers in my formative years, who created the foundations for my literary, academic and professional life in the ensuing years.

Particular thanks must also go to my American associate Dr David Liebers MD, whose command and application of the English language never ceases to amaze me. Finally, to the team at Greentech publishing who have provided the perfect balance of support, process and guidance in order to produce this first edition.

To my beautiful daughters.

Leah, Tamara, Billi and Ella

CONTENTS

Introduction.......... 9
GB Top 20 inventions and discoveries 12
Agriculture 14
Business innovation, retail and brands.......... 16
Clock making.......... 20
Communication – Audio-Visual 22
Criminology and Crime.......... 28
Cryptology 31
Digital and computing world 33
Discovery and English language 41
Emergency services and devices.......... 47
Engineering 49
Ethics, Philosophy, Religion, Politics.......... 64
Fabric manufacturing and process 72
Food and drink 77
Household and general innovations.......... 83
The Industrial revolution 94
Industry – process and innovation 98
Mathematics.......... 107
Military.......... 111
Music.......... 119
Photography and Cinematics 125
Publishing and authors.......... 128

Science 138

Astronomy 145

Biology 148

Chemistry 151

Medical innovation 155

Medicine 162

Physics 173

Sport 186

Television, cities and historic sites 196

Transport 203

Aviation 203

Rail 212

Road 217

Sea 230

Television, leisure and entertainment 234

Tradition and legend 240

Conclusion 245

About the author 249

Chronological index – by year 250

Great Britain
The mother of invention
Its contribution to the modern world

Introduction

A study by MITI – Japan's equivalent to Britain's Department of Trade & Industry — concluded that 54% of the world's most significant inventions of the last 100 years were British. Of the rest, 25% were American (USA) and 9% Japanese.

That figure becomes even more astonishing when set against Britain's modest geography, covering less than 1% of the world's landmass.

Yet Britain's contribution to the modern world cannot be fully captured by this statistic alone. Though impressive, it leaves out the global reach of the British system in human terms. The British university structure produces a culture of intellectual freedom and leadership requisite for creative output on this scale, and more world leaders have had their formative higher educational experiences in Britain than in any other country. According to the Higher Education Policy Institute (HEPI), a total of 58 current world leaders graduated from British universities.

Invention or innovation

British innovation and invention have been in abundance for centuries. But the word 'innovation' has only come into vogue in recent decades in order to describe the broader concept of how businesses or researchers adopt radically new methods to complement and advance traditional practices. This linguistic sleight of hand has happened in stealth, at the expense of the word 'invention,' which strikes the modern ear as primeval or coming from a past century, but not our own. The term 'invention' has quickly disappeared from the commercial lexicon.

The absence of invention in the modern world is easy to miss. On balance, most people see innovation as merely a modern rendering of the word invention. This is founded on the incorrect assumption that they have the same meaning. According to the Oxford Dictionary, invention is: 'the action of creating something (tangible) that has never been made before, or the process of creating something that has never been made before'. Whereas innovation is described in this way: 'to use a new idea or method'. What is true is that they both involve the act of creation.

Nevertheless, there is such a thing as pure invention that stands apart from pure innovation. Pure invention can take the following form: a new product is designed, produced, commercialised, and then sold to people, thus improving their lives. Invention often drives innovation when new inventions help drive changes in existing production practices.

This book is not intended to be the definitive book of British inventions. It is not intended to be an encyclopaedic or all-encompassing index of every British inventive achievement. Rather, it is an entryway. It is a point of reference that will be added to, refined and advanced over the coming years. It is principally about the people of Great Britain: the imagineers, creators, discoverers and inventors of this compact but ingenious nation.

The mother of Invention

Why write a book on invention in Britain? Certainly, there are larger, wealthier and more powerful nations. Great philosophers, scientists, inventors and thinkers can be found in all corners of the world. What is exceptional is the contrast between Great Britain's small size and it's standing as a world power. Britain is irrefutably 'Great' and possibly one of the greatest nations that the world has ever known. For such a small country, its achievements are colossal.

The aim of this book is to document a representative sample of these achievements in plain English, so as not to deter our younger readers with technical jargon or granular detail. My hope is that the reader will develop an understanding of the impact and scale of Great Britain's achievements. Ultimately, I have endeavoured to write a chronicle that can both serve as a reference tool for students and become a fixture on every coffee table.

During the compilation and writing of this book, I have drawn on a great number of sources, but I accept that there may be errors that are wholly of my own making. I will therefore be pleased to receive any corrections, great or small, along with any significant additions that merit inclusion in later publications. For this I would be eternally grateful. Please send to edits@greentechpublishing.com.

This book is a snapshot. Given the almost interminable inventory of British inventions, I made a choice to include the most memorable, distinguished and significant ones in recent history. If I have succeeded, then readers will be inspired to seek out more information and detail from further, more eminent sources. My thesis is simple: Great Britain is undeniably 'the mother of invention' in the modern world.

GB Top 20 inventions and discoveries

This top 20 of British inventions and discoveries is the definitive list from more than six hundred achievements documented in this book. Of course, it is with certainty that many will argue passionately as to what should be considered worthy of placing in the top twenty from such an illustrious collection of truly Great British invention and innovation. However, I have spent four years researching, compiling and publishing my first phase of Britain's disproportionate but truly colossal achievements versus the contribution from the rest of the world. I am therefore using author's prerogative to decide on a ground-breaking top twenty list of achievements that have truly contributed to positive change and progress in the modern world, to set the scene ahead.

1. English language
2. Penicillin
3. Hypodermic syringe
4. Vaccination
5. Electric motor
6. Electric generator
7. Television
8. Telephone
9. Steam engine
10. Programmable computer
11. World Wide Web
12. Passenger railway
13. Powered flight
14. Jet engine
15. Photography
16. Bicycle
17. Tin can
18. Stainless steel
19. Concrete
20. Float glass

"...to invent is to discover that we know not, and not to recover or resummon that which we already know".

Sir Francis Bacon

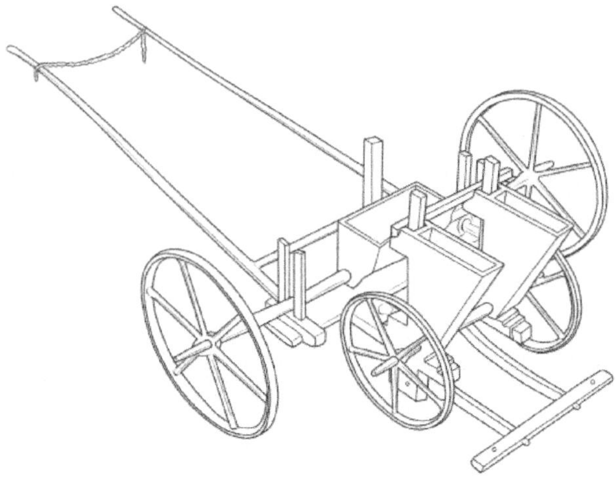

Seed drill - Jethro Tull 1701

Agriculture

Mechanical reaping machine 1828

Patrick Bell was a church minister born 1799, who studied divinities at the University of St Andrews. While working on his father's farm, Bell developed the reaping machine in an effort to speed up the hugely laborious manual task of harvesting. In 1828 his reaping machine, that was pushed and powered by horses, was used on other farms in his district to wide acclaim. Unfortunately for Bell, he never applied for a patent because it was his belief that his creation should benefit everybody. As a consequence, Bell did not make a financial gain on his invention as it enjoyed huge success across the world. In 1831, a patent was issued in the USA to William Manning for a reaper of very similar design.

Seed drill 1701

Born in London, inventor Jethro Tull's seed drill device gouged a straight rut into the ground at a pre-set depth, dropping seeds at regularly spaced intervals. It not only sped up the process of cultivation and crop planting, but also reduced wastage, helping to increase both output and life expectancy of the general population for the first time in history.

Threshing machine 1776

Andrew Meikle was a mechanical engineer who is recognized for inventing a contraption designed to remove the outer husks from grains of wheat, known as 'the Threshing Machine'. This was regarded as one of the foremost inventions in the world of agriculture and its concept is still used in the mechanised world of farming to this day.

Tractor Light commercial 1901

Daniel Albone was an inventor who developed the first commercial light farm tractor. Prior to Albone's invention, farm machinery was powered by horses or steam and they were generally very heavy and cumbersome and not at all portable. Steam propelled machines were too heavy to pull a plough in the soft mud. But by the 1890s, internal combustion engines had become more common and this spurred Albone to develop a high torque, lightweight petrol-powered engine for what he called his 'light-weight petrol powered general purpose agricultural vehicle'.

By 1901, Albone completed his first working prototype and by 1902 had filed a patent on 15th February. The term tractor did not come into use until a few years later. His tractor featured two huge rear wheels driven by the petrol engine and one wide small wheel up-front. The driver sat aloft on a single seat with all round open views. Albone's tractor was able to haul and power all farm machinery, previously the preserve of horses.

> "The discovery of agriculture was the first big step toward a civilised life".
>
> Arthur Keith

Mass production c1700

Business innovation, retail & brands

Chain stores World's first 1848

WH Smith & Son built the world's first chain stores. WH Smith began selling books and other goods at train stations belonging to the London & North-Western Railway, expanding to high street stores in 1905. There are now more than 1500 WH Smith stores in the company's portfolio.

Mail order 1834

Sir Pryce Pryce-Jones was an entrepreneur who created the world's first mail order business in 1834. Pryce-Jones capitalised on the local wool market and rather than take up farming like his peers, decided to start a business selling woollen flannels locally. When his business grew, he started to advertise in national newspapers and dispatched them by post, first in Great Britain and then overseas. This was the world's first mail order business.

Mass production C.1700

Mass production was first employed during the start of the industrial revolution in Great Britain in the Derwent Valley. Parts started to be manufactured elsewhere and then assembled in a factory on a facility known as a production or assembly line. This mass production resulted in a standardised product of mid to low quality. By its very nature, the work was very repetitive for the workforce. However, the assembly lines could be stopped as demand dictated and workers for the first time could be laid off, rather than stockpiling goods when demand is low, thus costs were greatly reduced in the process.

Mini skirt 1966

British fashion designer Mary Quant is generally acknowledged as the person who introduced the mini skirt and coined the phrase by producing short waist skirts that were set about six inches above the knee. Quant never took credit for the invention of the mini skirt but was instead recognised as the person who coined the phrase 'mini' for her short skirt designs.

Patent first documented granting of 1331

The security in sole ownership of a design, invention or innovation through the award of a patent is now taken for granted, but prior to 1331, any idea or invention could freely be copied and used by all. The first authenticated and documented royal grant of a patent was petitioned by and awarded to John Kempe and his company 'for the purpose of instructing the English in a new industry for products or services'. The description of the patent suggests that this may be an example of the world's first business consultancy company.

Rolex precision watches 1905

Rolex is recognised as a Swiss high-end watchmaker and ranks as 64th in the world global power brands. However, Rolex started its life in London as a precision watch making collaboration between Alfred Davis and Hans Wilsdorf. It later moved its base to Geneva, Switzerland and is now the world's largest luxury watch brand

Wealth creation 21st Century

Britain is the capital of the world for the super-rich, with London boasting more billionaires per head than anywhere else on Earth. Criticised by some, many of these wealth creators are at the heart of the booming UK economy and as their success and wealth grows, so too does the growth of job and wealth creation for the country as a whole. This is incredible for a country that occupies just 1% of the world's total landmass.

> "It is impossible to live without failing at something, unless you live so cautiously that you might as well not have lived at all, in which case you have failed by default."
>
> J. K. Rowling

Top 20 British retail brands (by value 2018)

1. Marks & Spencer
2. Boots Plc
3. Tesco
4. Asda
5. Next
6. Sainsbury's
7. Argos
8. Morrison
9. Waitrose
10. ASOS
11. B&Q
12. John Lewis
13. Debenhams
14. The Body Shop
15. Primark
16. Costa
17. Specsavers
18. Waterstones
19. Play (Rakuten Ichiba UK Ltd)
20. Wickes

TOP 5 British Banking Brands (by value 2018)

1. HSBC
2. Lloyds Banking Group
3. Royal Bank of Scotland
4. Barclays Bank
5. Standard Chartered Bank

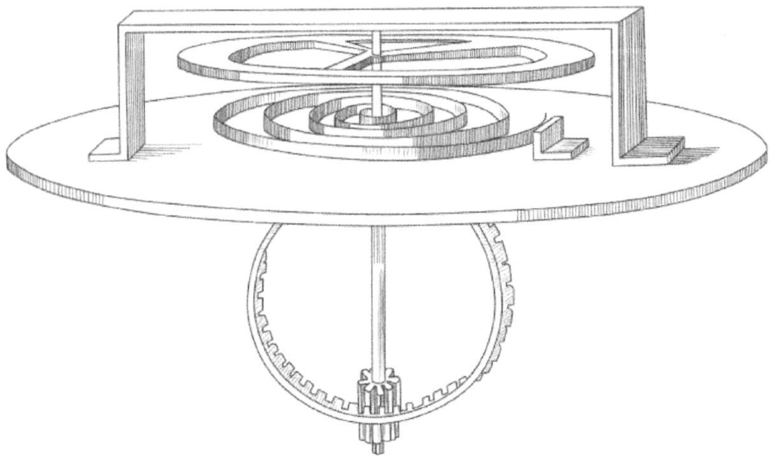

Balance spring 1660

Clock making

Balance spring 1660

Robert Hooke was an architect and philosopher, educated at Oxford University and born in the Isle of Wight in 1635. Hooke was responsible for many scientific and engineering inventions and theories, but it was his creation of the Balance Spring that revolutionized portable and pocket watches. His invention came about more than 15 years before Christiaan Huygens published his own work on an identical device in 1675.

Clock Oldest working clock 1386

Salisbury Cathedral is recognised as having the oldest working clock in existence. It is a large framed clock, without the common dial, located in the aisle of the great Cathedral. The clock belongs to a group of astronomical clocks from the 14th Century and uses the original mechanism for its operation.

Clock First electric 1841

Alexander Bain was an engineer and inventor who is credited with designing and making the world's first electric clock. His clock used a pendulum that maintained movement without a clockwork mechanism, by using electromagnetic pulses to maintain momentum. Bain was also famous for installing telegraph lines between Edinburgh and Glasgow.

Clock - First Atomic Clock 1955

Louis Essen was a scientist based at the National Physical Laboratory in Teddington London. He developed the first atomic clock that kept better time than a pendulum or quartz mechanism. Essen's clock was also considerably more constant than the Earth's rotation.

Wristwatch First automatic 1923

John Harwood is regarded as the pioneer of the automatic wristwatch. Harwood was a soldier and a respected watchmaker. After the Great War he founded the Harwood Watch Company. Harwood's aim was to devise a mechanism much better than existing clockwork watches. He noted that the presence of humidity and dust caused severe issues in accuracy of timekeeping in existing designs and so he set out to create a watch that would wind up from the inside of the watch, making it easier to seal. It was the sight of children playing on a seesaw that sparked Harwood's imagination and it set him on the path to inventing the world's first self-winding mechanism: the automatic watch.

> "By putting forward the hands of the clock you shall not advance the hour."
>
> Victor Hugo

Television 1925

Communication – Audio-Visual

Emergency calls 999 1937

Britain's 999 system is the world's oldest emergency call service and was introduced in London on 30th June 1937. Any phone call on the 999 number triggered a special light alarm system ensuring that the incoming 999 calls always took priority at the telephone operation centre. This ground-breaking emergency phone system was quickly adopted throughout the world, some using different numbering systems and is now very much part of the make-up of British life.

Fibre optics 1954

John Tyndall was a physicist who established that light signals could be sent through a curved stream of water, proving that light could be bowed. Tyndall later demonstrated the same effect by shining light into a pipe filled with water with bends in its length and as the water came out of the pipe into a tank, the water and light curved round in an arc. His discovery paved the way for glass optics that could be bent with minimal deterioration to light travel and output.

Greeting cards 1843

John Callcott Horsley was an artist who was commissioned by businessman Sir Henry Cole to design a Christmas card that he could send to his relatives, clients and friends. This was acknowledged as being the world's first greeting card.

Microphone 1878

David E Hughes was a scientist who is credited with inventing the world's first carbon microphone, essential to the development of the telephone. Hughes' invention was the basis for all future microphones still used today.

Postcodes or Zip codes 1857

The earliest system of postal districts was implemented in London and other large UK cities in 1857. In London, the system was further refined and expanded in 1917 to include numbered sub-areas, extending to other cities and large towns in 1934. The earlier district postcode systems were later incorporated into what is now the national postcode system.

Postmarks 1661

Henry Bishop was an inventor and the Postmaster General of England. Bishop conceived the world's first postmark for use on mail that became known as the 'Bishops Mark'. Bishop wanted to have a method of telling when a letter was received by the post service to ensure that the there was no delay in the dispatch of mail. The Bishops mark was also the predecessor to the world's first postage stamps.

Postage stamps 1680

British merchant William Dockwra devised the world's first pre-paid postage system, the London Penny Post. The Penny Post was the world's first inter-city mail delivery service at a fixed price and formed the basis for many other provincial Penny Post services in the years that followed.

Sign language 1680

George Dalgarno developed the world's first international linguistic sign language for deaf people. Dolgarno's method of sign language is used widely throughout Britain and the Commonwealth and still used in the USA.

SMS text message 1992

British mobile telecommunications engineer Neil Papworth was working for telecom company Vodafone, when he became the world's first person to send an SMS or text message in 1992. Papworth sent the text from his computer to Richard Jarvis, a field engineer, on his Orbital 901 mobile phone, with the message 'Happy Christmas'. Texting is now the single most popular method to stay in contact. According to British regulator OFCOM, 12-15-year-olds now send an average of 193 texts every week, with the average person sending 50 per week.

Stereo sound 1931

Alan Dower Blumlein patented a technique for playing stereo sound with a single pick up and stylus. Later employed by EMI in 1933.

Telegraph Electric 1837

Sir Charles Wheatstone & William Cooke filed a patent for the electric telegraph in 1837, which was granted the same year. The first fully functioning telegraph ran between West Drayton and Paddington railway stations in London, but the revolutionary invention's huge benefits were not capitalised. That was until 1845 on New Year's Day, when the electric telegraph helped seize murderer John Tawell.

It caused a sensation and from this point on, Telegraph systems were soon spreading globally throughout the civilised world.

Telephone 1876

Alexander Graham Bell applied for a patent for the telephone in 1876; later granted by the US patent office. British inventor, Bell patented his telephone model literally hours before a rival inventor. It evolved largely due to a discovery using a thin metal sheet that vibrated in an electromagnetic field. This generated an electrical waveform equivalent to the vibration. A demonstration of this invention was first publicly displayed in 1876 at the Centennial Exhibition in Philadelphia.

Television Cathode Ray Tube 1908

British engineer, Alan Archibald Campbell Swinton, published the world's first available article that proposed an electronic television using a cathode ray tube.

Television 1925

John Logie Baird submitted a patent for a mechanical television system in 1925. He later demonstrated his design in 1926; the first to transmit fully moving pictures. Unfortunately his mechanical system was futile, with the advent of a competing alternative, offering a noticeable superior picture quality than Baird's system.

Television Transatlantic transmission 1928

John Logie Baird made the first transatlantic television transmission by land cable to an international radio transmitting and receiving base, using his electro mechanical system.

Television colour 1928

John Logie Baird made the first colour television transmission using scanning discs at the receiving and transmitting terminations. He used three spirals of apertures with a different primary colour filter in each one. He installed three lights at the receiving termination point and used a commutator to oscillate the illumination. This was to be the world's first colour Television Transmission.

Television Pocket TV 1983

Electronic engineer, scientist and inventor Sir Clive Sinclair designed and developed the world's first pocket TV called the TV80. The development costs were high and sales demand was low and although the TV80 was a technological marvel, Sinclair's marketing team appeared to be failing in their target market, or perhaps the TV80 was just too far ahead of its time.

Transatlantic cable 1850

The world's first undersea cable laid in the English Channel was installed between England and France, followed by a transatlantic cable in 1857 led by telegraph engineer, John Watkins Brett. He had a vision that telegraph cabling would create a new world of instantaneous communication. In 1847 the French government approved Brett's application to lay a submarine cable between Dover and Calais. Only a handful of calls were made before a fisherman struck and severed the cable, rendering it inoperable. In 1851, Brett founded the Submarine Company and by 1856 had raised enough finance to fund the Atlantic Telegraph Company, who laid the first transatlantic cable, but due to poor quality and installation techniques it had failed within the first few weeks of operation. Unfortunately, Brett died before witnessing the first successful cable installation later that year in 1866.

Video recorder 1928

John Logie Baird filed the first patent for video recording, to record moving images on wax discs. His device was a by-product of his ground-breaking television invention and he named this new device Phonovision. It was constructed by the use of a large Nipkow disk attached mechanically to a standard 78rpm record cutting lathe. This produced a permanent record in 30-line analogue video resolution that could be played back through his television at 30-line quality. There was no further development or commercialisation.

Whistles 1883

Joseph Hudson was a toolmaker from Birmingham who converted his tiny washroom in his rented flat into a makeshift workshop. He worked on anything he could to supplement his meagre income, even repairing shoes at weekends. He crafted his first whistle in this workshop and later formed a company with his brother James, called Acme Whistles. It is still the world's most famous and largest producer of whistles today, still based in the Jewellery quarter in the City of Birmingham.

Finger printing 1684

Criminology and crime

Broadmoor Hospital 1863

The world's first dedicated asylum for criminal lunatics is based in the Village of Crowthorne, Berkshire, not far from Windsor Castle, the Queens second royal residence. It is reputed to have the highest level of security for any public building in the country. Broadmoor Hospital also boasts tougher fortification than any British top category, high security prison.

Arsenic Test 1836

Chemist, James Marsh developed the Marsh test. It is a highly accurate method for the detection of arsenic, particularly useful in forensic toxicology to detect arsenic as a poison. Arsenic used to be a highly successful and undetectable poison prior to the Marsh test and was almost untraceable in the body.

DNA Fingerprinting 1984

After seven intensive years of research, Alan Jeffreys, a geneticist from Leicester University, developed a technique for distinguishing one person's DNA from another. It was on September 10th 1984 when he suddenly realised differences and similarities in the DNA between his family members and one of his technicians at his laboratory. It was then that he realised the huge potential of his DNA profiling for use in police forensic work and soon worked with police on the murders of local girls Dawn Ashworth and Lynda Mann, both raped and murdered in Narborough, Leicestershire.

Tests proved that the DNA from both semen samples were from the same man and not from the prime suspect who had previously been arrested because he had the same blood group.

It was proved beyond doubt that this man was innocent. Jeffreys subsequently filed a patent for the use of DNA fingerprinting to identify individuals following real life tests over many years of research, concluding with tests on the rapist and murderer, Colin Pitchfork, proving beyond doubt that he was responsible for this heinous crime in Narborough, and a patent was subsequently granted in 1986. This technique has now been used to catch thousands of criminals worldwide through conclusive DNA evidence.

DNA database 1995

On April 10th, 1995, the world's first DNA database was launched in Britain and was cited as being the biggest breakthrough in the fight against crime since fingerprints, by the then Home Secretary Michael Howard.

Face recognition 2008

A British joint venture between BAE and Omni Perception has improved suspect identification on CCTV footage and developed lip movement with speech pattern recognition technology. This new combination of speech and lip-reading recognition, offers police and other law enforcement agencies a means to secure valuable evidence about a suspect's identity from CCTV, far quicker and more accurately than has ever been possible before.

Fingerprinting 1684

Scientist Nehemiah Grew is recognised as the first person to understanding and document the detail and relationship of fingerprints. Grew used a microscope to study and research the ridges found on palms and fingers and published his first paper on the subject in 1684.

Robert Melias 1987

Robert Melias, sentenced for 8 years in prison at Bristol Crown Court for Rape, notoriously became the first person in the world to be convicted of a crime using the evidence of DNA.

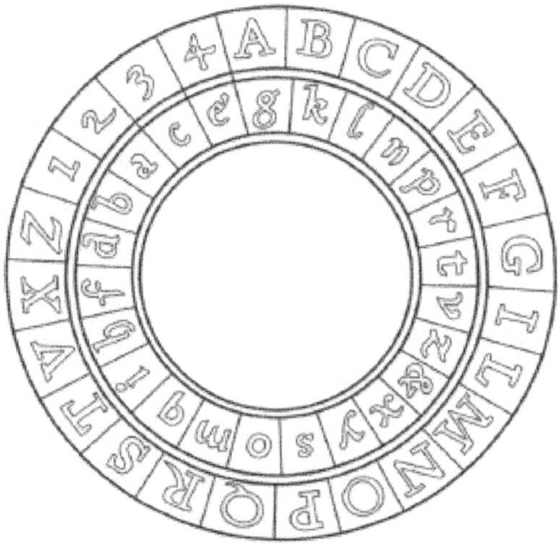

Early Cypher

Cryptology

Cipher Bacons cipher 1605

Francis Bacon devised a technique to hide a secret message that was hidden in the body of text, rather than just a cipher. This method of encryption was known as Bacons Cipher or Baconian Cipher.

Cipher Playfair Cipher 1854

Charles Wheatstone invented the world's first manual symmetric encryption cipher but bore the name 'Playfair Cipher' because Lord Playfair was the promoter of this unique device. The cipher encrypted pairs of letters, rather than earlier ciphers that encrypted single letters. This made its messages much harder to break, since the method of frequency analysis used on simpler ciphers did not work well. The device was used in the Second Boer War, WWI and WWII because it was quick to set up and use and fairly reliable.

RSA cipher 1973

Clifford Cocks is a mathematician and cryptographer, who, while based at GCHQ in Cheltenham, discovered the encryption algorithm now commonly used and known as RSA. His algorithm was created more than three years prior to being independently developed by Rivest, Shamir and Adleman at MIT in the USA during 1977. At the time he was not recognised for his creation because working for GCHQ meant that his work was classified information and therefore not available in the public domain.

Bletchley park 1939-1945

More than 12,000 men and women decoded the German Lorenz and Enigma Ciphers on this Buckinghamshire site, during the second world war. These remarkable people also broke down Italian and Japanese ciphers and codes, by implementing pioneering mathematical analysis, assisted by two huge computing machines. The machines were designed and developed by teams led by Alan Turing: the electronic Colussus that was designed by Tommy Flowers and the electro-mechanical Bombe developed with Gordon Welchman. The huge team achievements at Bletchley Park, significantly shortened the war, saving many lives in the process.

> "Cryptography is typically bypassed, not penetrated"
>
> Adi Shamir

Digital and computing world

ATM (Cash machine) PIN 1966

James Goodfellow patented the Personal Identification Number (PIN) technology and is generally acknowledged as one of the most significant co-inventors of the ATM (Automatic Teller Machine). Goodfellow was a British Engineer and given the project to develop an automatic cash dispenser in 1965. His system allowed a machine-readable card to be used. The cards were encrypted for security through the use of magnetic strips allied to a four-digit PIN number, accessed by the ATM keypad.

ATM (Cash machine) 1967

The world's first ATM or cash machine, was established at a branch of Barclays Bank in Enfield, North London. John Shepherd Barron was the chief inventor and worked for the world's largest bank note printing firm, De La Rue. The first person in the world to use the ATM was comedy actor, Reg Varney, star of the TV sitcom 'On the buses'. At the time, the ATMs were known as DACs (De La Rue Automatic Cash System). Shepherd Barron was the General Manager for De La Rue Instruments who made the world's first ATM's.

ARG Electrically powered analogue computer 1906

Arthur Pollen was a naval affairs writer and through his huge naval experience recognised the need for a computer-based fire control system on war ships, due to the wholly inaccurate method of manual firing of the day. He set about designing a suitable system in the early 1900s and later produced the first version of what is now recognised as the first computerised fire control system. It took account of the tide and its effect on the ship by using a tidal analyser.

Later in more advanced versions; Pollen integrated a gyroscope to take account of the ships yaw and used a graph plotter to capture the target's position and relative motion. For the first time ever, the British Navy had automatic assistance in range and firing, producing startlingly accurate results.

ARM Universal mobile device processor 1978

Mobile phone processor designer, Arm Holdings based in Cambridge, has grown from making one of the world's first home and school computers, the Acorn, to being one of the world's most important semi-conductor designers. The ARM processor provides the heart to more than 95% of smart-phones and tablets globally; including iPad, iPhone, Microsoft, Samsung, Sony and LG. Significantly, ARM holdings is not a manufacturer, but a designer and licensor of its power processors. In 2008, the ten billionth ARM based processor was shipped.

Bar codes commercial application 1970

Bar coding was first used commercially in 1966, but it was not until British company Plessey developed a universally acceptable system, that its use became widespread and internationally accepted. The first company to universally adopt this system of stock control was the British food retailer J Sainsbury, used for every item of stock in their portfolio.

Biometrics Iris recognition 1949

Ophthalmologist James Doggart wrote 'just as every human has different fingerprints, so does the minute architecture of the iris exhibit variations in every subject examined. Its features represent a series of variable factors whose conceivable permutations and combinations are almost infinite'. This was the first time a scientific paper had been written and published on this subject.

Biometrics Iris recognition algorithms 1994

Ophthalmologist John Daugman developed and patented the first algorithms to recognise the individual iris of a human eye. Daugman's algorithms became extensively licensed by a series of different commercial organisations. Today, after many improvements, Daugman's algorithms remain the foundation of all commercial employment of iris recognition technology.

Calculator Pocket 1972

The world's first 'slimline' pocket calculator was launched in 1972 at the huge price of £79.95 Plus VAT. It was designed by inventor and Scientist, Clive Sinclair and manufactured by Sinclair Radionics. In 1973 a new version was launched called the Sinclair Executive Memory. Sinclair went on to make many more breakthroughs, such as the ZX Spectrum, one of the first hand held computers and the Sinclair C5 3-wheel electric transporter.

Computer World's first computer 1820

Charles Babbage was a renowned philosopher, mathematician, engineer and inventor. Babbage is credited with creating the world's first mechanical computer, the 'Difference Engine'. Babbage designed this machine to automatically calculate polynomial functions. The name 'Difference' derives from the system of divided differences; a method to tabulate functions using a set of polynomial coefficients, both trigonometric and logarithmic functions.

Computer programme First programmer 1843

Mathematician Ada Lovelace is recognised as being the world's first computer programmer, by writing instructions to drive the first proper program on one of inventor Charles Babbage's early computers, the Analytical Engine. Lovelace wrote and published a paper on the subject of programming and spent many years perfecting and refining her newfound art.

Computer First stored programme computer 1948

The SSEM was developed and built by computer scientists Frederic C Williams, Tom Kilburn and Geoff Tootill, all from the Victoria University of Manchester. The Manchester Small Scale Experimental Machine (SSEM) was not intended to be a mass market computer, instead designed as an early test bed for the 'Williams tube', a primitive form of computer memory. The SSEM was the first working machine to enclose all the fundamental elements of a modern computer.

Computer Laptop. 1979

Industrial designer, Bill Moggridge, designed the Grid Compass. It was the world's first commercial laptop computer with clamshell design and flat panel screen. Moggridge is attributed to designing a 'computer in a suitcase' and when it was released for sale, it had the staggering price of £5,091. The Grid Compass had a magnesium case with a yellow on black keyboard and was eventually used by the US military, later having the distinction of being used on the Space Shuttle.

Computer Portable. 1981

Adam Osbourne, a computer scientist is credited with developing the world's first commercial portable microcomputer, not to be confused with a laptop. US industrial Designer Lee Felenstein designed the Kaypro aesthetics. Osbourne planned to break the high price of computers. The KayPro design was loosely based on the Xerox Note Taker. It was intended to be both robust and portable, clad in an ABS (Acrylonitrile butadiene styrene) case and hidden handle. Unfortunately, it was about the weight and size of an accordion and had the looks of a WW1 telecommunication radio. At its peak, the Osbourne 1 sold 10,000 units a month.

Computer Handheld 1984

Computing pioneers PSION developed the Psion Organiser. The Psion Organiser is regarded as being the world's first handheld computer and was launched as the PSION 1, under the brand name PSION Organiser. It was the world's first useable PDA (Portable Digital Assistant) and was characterised by its sliding keyboard cover and LCD display. It weighed 225 grams and measured 142 x 78 x 29.3 mm, equipped with fully static RAM with impressive battery life for its time. It was originally priced at £99. In 1986, PSION introduced an improved version called the PSION 2.

Computer Electronic programmable 1943

Tommy Flowers was an ingenious British Post office engineer by profession. But it was the war effort and great need for powerful encryption of coded communications from the German Lorenz Cipher that gave Flowers the inspiration to create the world's first electronic programmable computer – The Colossus. Far more complex and powerful than the more famous Enigma, it was the world's first truly programmable electronic computer, and used more than 1500 valves or vacuum tubes as they were then known. Unfortunately, such was the secrecy and the advanced power and technology of the Colossus, that it was only ever used for Military encryption and remained a secret until long after the Second World War was over. Unfortunately, every Colossus ever made was dismantled.

Diode rectifier tube Radio Valve 1904

Sir John Ambrose Fleming was a renowned scientist who developed the 'Fleming Valve' or oscillation Valve. It was the world's first thermionic or diode valve, effectively the electronic equivalent to a one-way water valve. It was the first thermionic diode and was used extensively in later years in radio receivers and also as a low current rectifier in power supply units.

Fax machine 1843

Engineer and inventor Alexander Bain was the first person in the world to file a patent for the facsimile machine (FAX), but with the exceptions of a few improvements in the years ahead did not develop it. This was an opportunity for Frederick Bakewell who beat Bain with a demonstrable model of a fax machine that he called the 'image telegraph'.

Fax machine First public display 1851

Physicist Fredrick Bakewell patented and demonstrated the world's first public display of a working fax machine transmission in 1851 at London's World Fair. Bakewell's 'Image Telegraph' encompassed many elements that are built into today's modern fax machines, replacing some of the crude mechanics of Bain's design and replacing them with rotating cylinders.

Hard drives Computer 1997

Professor Stuart Parkin developed high capacity computer data readers. These devices allowed more information to be stored on each disc platter than previous incarnations and this invention has now laid firm foundations in data storage protocol and development, enabling the storage needed for huge organisations such as Amazon, Google, Facebook and other online services.

HTTP URLs and HTML 1989

Famed for his development of the World Wide Web, computer scientist and Oxford educated Tim Berners-Lee also simultaneously developed URLs, HTTP and HTML, later refined as web technology rapidly developed.

Integrated circuit or Microchip 1952

G.W.A. Dummer was an electronics engineer from Hull, who was the first person to create and construct a prototype of the Integrated Circuit in 1952, better known as the Microchip. Through his work at the Telecommunications Research Centre, he believed that it was feasible to build multiple circuits on and into silicon or a similar substance. While at a conference in Washington DC, he presented his work. Six years later, Jack Kilby of Texas Instruments was awarded a patent for essentially the same design raising many questions about the legitimacy of the award. Foyle is known as the 'father' or 'prophet of the integrated circuit'.

iPod® Origins 1979

Electronics engineer Kane Kramer, patented the revolutionary IXI player, a device that stored music on a digital chip. Holding just three and a half minutes of music, barely enough for a full single. His patents ran out in 1988 and because he couldn't raise the £60,000 needed to renew them, the Apple® iPod was revealed to the world in 2001.

iPod® 2001

Designed by British born Apple chief designer Sir Jonathan Ive, who is one of the senior directors of the giant computer company; Apple®. The iPod® has since sold more than 170 million units. The Daily Telegraph recognised Jonathan Ive in 2008, as being the most influential Briton living in America today.

Packet switching 1965

Donald Davies is an inventor who developed the Internet protocol for sending and receiving information in parcels or packets of information. Now taken for granted, the way information or data is packed and transmitted on both the Internet and many mobile phone systems globally is the de-facto standard for all operating systems.

RAM Random Access Memory 1918

William Eccles and F.W. Jordan were physicists who developed the 'flip-flop' circuit in 1918. This methodology became the foundation of electronic memory in computers still used worldwide today.

Trackball computer pointing device 1952

The trackball was developed in project DATAR, jointly developed between British engineer Kenyon Taylor and the Canadian Navy. Its purpose was to front a new battlefield computerised information system and this was the precursor to the computer mouse.

Web browser 1990

Computer scientist Tim Berners-Lee invented the world's first web browser. It was then called WorldWideWeb (all one word), but was later renamed NEXUS. Berners-Lee used a neXT computer and pioneered the use and terminology of 'hypertext' for information sharing. He was also responsible for creating the world's first web server. Since his creation, the web and web browsers are seen as one by most users.

WWW World Wide Web 1991

The World Wide Web was developed and invented by computer scientist Sir Tim Berners-Lee. He created his first server in 1990 and on August 6th, 1991, his web went live with its first page describing how to search and set up a web site. Berners-Lee took no fee for his revolutionary invention, which he gave to the world completely free. This is not to be confused with the Internet, which is a network of linked computers

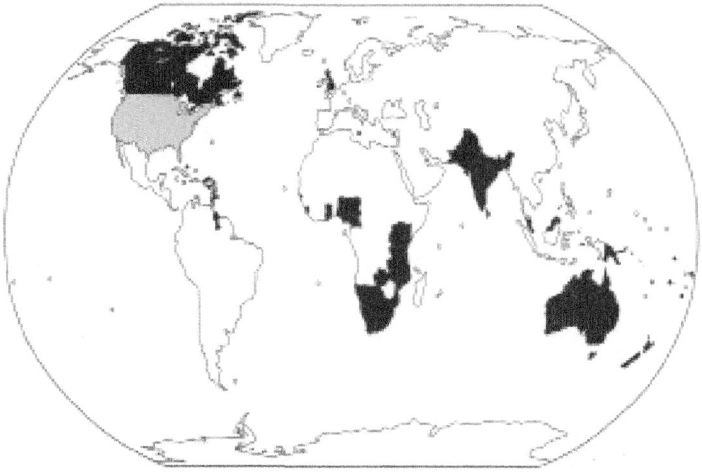

Commonwealth countries (former GB territory USA in grey)

Discovery and English language

America Discovery of what is now the USA 1170

Prince Madog Ab Owain from Wales landed in the present-day town of Mobile, Alabama. Almost 350 years before Christopher Columbus landed in the Caribbean and long before the first documented European, Spanish Conquistador Juan Prince de Leon arrived on the mainland of what is now known as the USA on April 2nd 1513.

However controversial this entry may be, there is no doubt that Great Britain played a huge part in establishing the USA, the world's most formidable super power in the guise we see today. The USA was created when established British Colonies in North America declared independence from what was then known as the Kingdom of Great Britain. The motivation that led to this break-up was to do with taxation and what was seen as continually rising duty for no apparent benefit to the people of the American colonies. The War of Independence lasted from 1775 to 1783 with the American colonial forces, aided by French, Dutch and Spanish forces, claiming ultimate victory. The reality, however controversial, remains though, that the victors were British subjects until that day in 1783. Therefore, on that basis, it cannot be disputed that the USA was, by its very establishment, created by the British

Australia Formation. 1770-1788

Captain James Cook was an explorer, cartographer and navigator in the Royal Navy. He produced detailed maps of Newfoundland before making three voyages to the Pacific Ocean, in which he made his first contact with the Hawaiian Islands and the eastern coastline of Australia. He also recorded the world's first circumnavigation of what is now New Zealand. As he advanced his voyages of discovery, he surveyed, mapped and started to name features. He recorded and started to name islands and coastlines of these new far-flung lands for the first time

Cook subsequently charted the Australian east coast in his ship HM Barque Endeavour. Under instruction from King George III of England, Cook claimed the Australian east coast in 1770. He named Eastern Australia as 'New South Wales'. Cook was sadly killed during his third Pacific trip to Hawaii during a fight with locals. His legacy of geography, cartography and scientific knowledge was to influence his successors into the 20th century. Then in 1788, Captain Arthur Phillips and his first fleet, with 11 ships and 1350 people, arrived at what is now known as Botany Bay, moving quickly on to Port Jackson because Botany Bay was considered unsuitable for settlement. Captain Phillips had instructions to establish the first British colony in Australia in the Sydney Cove area, now the harbour of one of Australia's greatest cities; Sydney.

The Commonwealth 1583

The original British Empire, between 1583 and 1783, developed when Walter Raleigh and Humphrey Gilbert were granted permission by Queen Elizabeth I to establish colonies in North America.

The second phase of the British Empire between 1784 and

1815 majored on controlling governments and infrastructure of the colonies and the increase of trade. India was a good example, where in the latter part of the 18th century, the ruling Mughals relinquished power to the British.

The third phase of the British Empire between 1815 and 1914 was known as the Imperial Century. It was in this decade that Britain controlled more than 10 million square miles of the World, inhabited by more than 400 million people. Britain was such a global force, that even economies that it did not govern, such as China, were still controlled from London.

At the start of the 20th century, there was a latent wave of countries wishing to become independent from the direct rule of Britain, but liked the idea of still remaining within the family of colonial Britain. In fact, as far back as 1814, during a visit to Australia, Lord Rosebery first described the evolving nature of the British Empire by calling it a 'Commonwealth of Nations'. Then in 1921, the term 'British Commonwealth' was first officially used by statute, to replace 'British Empire' in the wording of the 'Anglo-Irish Treaty' of the same year.

In 1926 at the Imperial Conference, Britain agreed that it and all its dominions were of equal status in every aspect, and that all were united in their allegiance to the Crown. From this point on, this great group became known as 'The British Commonwealth Nations'.

After WWII, the former British Empire was slowly dismantled, leaving only 14 wholly dependent territories within the family of the British Commonwealth, and in 1949, the term 'British' was removed from its name, leaving 'The Commonwealth' to truly reflect its changing nature and role in the modern world.

Today, The Commonwealth includes some countries that are newly formed by modern standards or who were not previously part of the old British Empire. The current membership of Commonwealth countries (2014), totals 53 nations in all. The membership includes countries spanning the Americas, Africa, Asia, the Pacific and Europe. The countries range from some of the largest countries in the world to some of the smallest countries. They also vary from some of the Poorest to the richest countries. All members of the Commonwealth have to pledge to its principles and values as written in 'The Commonwealth Charter'.

Regardless of economic status or size, all member countries have an equal say and vote and this protocol, ensures that even the smallest countries have an equal say in developing the Commonwealth. The leaders of each member country meet up bi-annually to discuss the issues affecting both the wider world and The Commonwealth, with the one aim of shaping its priorities and policies.

The Commonwealth members 2014

Africa

- Botswana
- Cameroon
- Ghana
- Kenya
- Lesotho
- Malawi
- Mauritius
- Mozambique
- Namibia
- Nigeria
- Rwanda
- Seychelles
- Sierra Leone
- South Africa
- Swaziland
- Uganda
- United Republic of Tanzania

Caribbean and Americas

- Antigua and Barbuda
- Bahamas
- Barbados
- Belize
- Canada
- Dominica
- Grenada
- Guyana
- Jamaica
- Saint Lucia
- St Kitts and Nevis
- St Vincent and The Grenadines
- Trinidad and Tobago

Asia
- Bangladesh
- Brunei Darussalam
- India
- Malaysia
- Pakistan
- Singapore
- Sri Lanka

Europe
- Cyprus
- Malta
- United Kingdom

Pacific
- Australia
- Fiji
- Kiribati
- Nauru
- New Zealand
- Papua New Guinea
- Samoa
- Solomon Islands
- Tonga
- Tuvalu
- Vanuatu

English language

English is spoken as a main language by a fifth of the Earth's population. The story of English as a language is one of struggle and survival against all odds, over many centuries encapsulating its development and evolution. The English language has been shaped and influenced over the years by Romans and Italian scholars who introduced Latin; the German influence derived from their barbaric tribes; Norse influence from the Viking invasions and Normans from France who introduced many French sayings that we now take for granted. With all these external pressures, the English language has had to exert a huge amount of resistance. More people, simply by virtue of China's huge population, speak Chinese Mandarin. However, English is now central to science, business, Internet, international air traffic control, world politics and diplomacy worldwide.

New Zealand Discovery and settlement 1769

The establishment of people in New Zealand dates back at least 700 years when Polynesians settled and started forming what is now recognised as the Maori culture.

Then in the early 17th century, the Dutch, fresh from victory over the Spanish, grabbed most of the East Indies from the Portuguese and so keen to develop the great continent of the south, that many people believed lay between Cape Horn and Australia, commenced a voyage of discovery. They first discovered Tasmania, named after the Dutch captain Tasman and encountered fierce opposition from the local Maori warriors. He quickly cast off and headed north; unaware that he was just a few hundred miles off the 'great southern continent'.

It was left to British naval commander and explorer James Cook to finally discover what is now known as New Zealand in 1769. Then in 1840 after many years of negotiation between the British government and the Native Maoris, the Treaty of Waitangi was signed welcoming New Zealand into the British Empire, giving Maoris the same rights as British subjects.

> "One doesn't discover new lands without consenting to lose sight of the shore for a very long time"
>
> Andre Gide

Emergency services & devices

Fire brigade 1824

Fire Master James Braidwood established the world's first municipal fire brigade in Edinburgh and is recognised as being the founder and developer of the modern metropolitan fire service.

Police World's oldest police force 1829

The Metropolitan Police Act of 1829 formed the basis for London's Metropolitan police force that was founded in the same year. The 'Met' as it is now affectionately known, was a merger of the Bow Street Horse Patrol and the River Thames Marine Police Force dating back to 1805. Still in operation today, the Metropolitan Police force is the world's oldest continuously operating police force in the world.

Fire extinguisher Modern portable 1816

Inventor George William Manby developed the world's first pressurised cylinder fire extinguisher. Manby witnessed firemen struggling to extinguish a blaze at a house fire in Edinburgh. This prompted him to design a portable copper made cylinder holding about four gallons of potassium carbonate, dispersed by air pre-compressed with a regulator to control the output when operated.

Lifeboat World's first 1774

Engineer Lionel Lukin submitted a patent for the world's first non-submersible lifeboat. He modified and patented a 20ft Norwegian yawl and fitted it with watertight cork filled chambers for extra buoyancy and then fitted a cast iron keel to keep the boat upright.

Lifeboat station 1776

The world's first lifeboat station opened in Bamburgh, Northumberland, with a LUKIN lifeboat. Its claim to fame is that it was the scene of the world's first lifeboat launch. The original lifeboat house has since been converted into a quaint self-catering cottage.

Water sprinkler 1723

The world's first water sprinkler system was designed and developed by chemist, Ambrose Godfrey. The device consisted of a barrel of water mixed with chemical dry powder of Godfrey's design. Included in the barrel was a sealed tin containing gunpowder with an external fuse. The patent filed by Godfrey and granted in 1723, informs that the fuse was lit by the flames of the fire, then the gunpowder exploded, splitting the barrel, thereby splashing the water and chemical contents around to extinguish the flames. It proved to be completely successful in at least one test.

> "Emergency services... a good day is when everybody gets to go home..."
>
> Anon

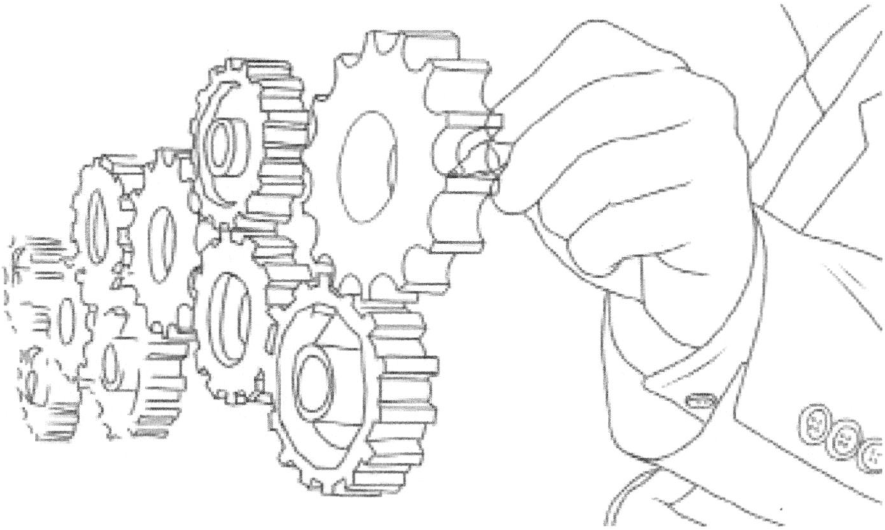

Engineering

Adjustable spanner or wrench 1830

Edward Budding was an engineer from Stroud in Gloucestershire who invented the world's first adjustable spanner. Inventor and engineer Richard Clyburn later improved the design in 1842.

Aquarium Design and construction 1853

Phillip Gosse was a naturalist, who conceived and developed the world's first large-scale aquarium. His creation was not only used to display aquatic species, but he also designed the huge tank to be used for plants and reptiles. Thus, in the process, Gosse opened the world's first public display aquarium, completed and open to the public in 1853 at Regent's Park, London.

Ball bearing 1794

Phillip Vaughan took out the first patent on ball bearings in this year. Up until this pioneering invention, moving parts on machinery tended to wear out very quickly as they rubbed against each other, but Vaughan came up with the notion of using solid steel balls secured in a groove to reduce friction and take the stress and strain out of moving parts. This great invention is now used in a plethora of products from skateboards to wheels on cars and from lorries to drive shafts on petrol and electric vehicles.

Bank notes World's largest manufacturer

British company De La Rue has been established for more than 200 years and is the world's largest commercial manufacturer of Bank Notes. The British company is the official bank note producer for more than 150 nations worldwide and is also responsible for officially producing the passports for more than 80 countries globally.

Bridge Box Girder 1848

Engineer Robert Stephenson built the world's first box girder bridge, a railway bridge built between Llandudno and the town of Conwy in Wales. The Conwy wrought iron bridge was built to a design by engineer and bridge designer William Fairbairn and constructed by British inventor and engineer Robert Stephenson. The bridge was completed in 1848 but not officially opened until 1849. Because this was the world's first ever bridge to be built in this form, the designers of the concept, Fairbairn and Stephenson, needed to carry out intensive proving of prototypes to demonstrate that it could safely carry the huge weight of railway locomotives. Aside from reinforcing columns being introduced under the bridge in later years due to the advent of even heavier and more frequent locomotives, the bridge remains intact since its original construction.

Bridge Concrete 1877

The world's first concrete bridge was built in Seaton, Devon, over the River Axe. British engineer, Philip Brannan built the triple span concrete arch structure in 1877, with the middle arch spanning 50 feet. The bridge still stands to this day.

Bridge Iron 1779

Engineer Abraham Darby was a pioneer in the process smelting local iron ore by using coke from local coal. He knew that industrial development was being held back in his area by the lack of a bridge over the River Severn. The main issue was that the bridge had to be single span due to the huge amount of barge traffic on the river. An Iron bridge was originally suggested and designed by Thomas Prichard, an architect from Shrewsbury, but he died just as work commenced. Abraham Darby III who had the resource and the knowledge to build such a ground-breaking bridge design, then adopted the project. The work commenced and after more than 380 tons of iron was transported from Darby's smelting works to build the 33yard (30m) span bridge, it officially opened on New Year's Day 1781.

Bridge Railway 1727

Engineer Ralph Wood built the world's first railway bridge near Stanley, County Durham, known as the Causey Arch; it is the world's oldest surviving single arch railway bridge. The Causey Arch Bridge was funded by the 'Grand Allies', a local consortium of coal-owners and designed and built by local stonemason Ralph Wood.

Documentation from archives confirm that the bridge cost £12,000 to build and was used to take coal from the pits to the River Tyne and then back via horse drawn wagons on a double track. It was the longest single span arch bridge in the country when it was built. The Causey Arch Bridge was reinforced and restored in the 1980s and now forms part of a series of scenic walking routes in the area around the Causey Burn.

Bridge Tallest. GB designed - French built 2001

Why a French Bridge? Well, the Millau Viaduct in France is designed by architect Norman Foster and engineered by structural engineer Michel Virloeux (Fr). It is the tallest bridge in the world with one of the supporting piers being 343 metres above the bridges base.

Mast pier 2, at 343m is higher than the Eiffel tower (324m), making it France's tallest structure.

Bridge Tilting 2001

British architects Wilkinson Eyre designed the world's first tilting bridge. The Gateshead Millennium Bridge is designed for cyclists and pedestrians only and traverses the River Tyne between the quayside of Newcastle-upon-Tyne and Gateshead's Southbank Quays quarter. The bridge won a design award and is sometimes referred to as the 'blinking eye bridge' due to its tilting method and shape.

Caterpillar tracks Patent 1770

Richard Edgeworth was an inventor and writer, born in Bath. Among his many inventions, Edgeworth filed a patent for what is now commonly known as caterpillar tracks. He submitted a design for a steam engine that ran on an 'endless railway system'. Flat steel planks were attached to the main wheels of the steam engine to ensure operation on soft ground, but in reality, Edgeworth's system was not successful until 1846, when James Boydell submitted a more feasible system.

Caterpillar tracks First commercial use 1903

The world's first vehicle to utilise caterpillar tracks was named a 'crawler tractor' and was built by David Roberts of Ruston Hornsby & Sons.

Channel Tunnel Anglo-French 1994

The Channel Tunnel is the longest undersea rail tunnel in the world (actually three tunnels), at 34 miles long with 24 miles being under the sea and often called the Chunnel or Euro Tunnel. The tunnel lies under the water of the English Channel and connects Great Britain with mainland France; completed in 1994 it is considered to be possibly the greatest engineering achievement of the 20th Century.

However, for centuries before its completion, both the French and the British had various proposals presented, but through cost or security implications, nothing materially happened.

In fact, there still remains evidence of an early British attempt to tunnel from Dover to France in the 19th Century. In 1876 an Anglo-French agreement was made to build a railway tunnel under the sea and link our two great nations under the guise of the Submarine Railway Company. An access shaft was drilled in 1880 and a seven-foot diameter test-tunnel started below Abbot's Cliff midway between Dover and Folkestone. What made this venture different from previous attempts was that an ingenious boring machine was developed and made by Captain Thomas English to bore the tunnel. It was 33 feet in length and could tunnel almost half a mile a month. By 1881, after more than 800 feet had been drilled towards France, with the machine proven, work was then diverted closer to Dover and accessed by a shaft of 150-foot depth below the sea bed. But by 1882, the government stopped all boring, for fear of the possible military repercussions of this land-link to France, resulting in all shafts being backfilled and mothballed.

It wasn't until 1984 that both Prime Minister Margaret Thatcher and French President Mitterrand jointly agreed that the Channel link would be mutually beneficial. Neither country had the financial means to fund such a project, so a contest was organised for private consortiums to build and run the tunnel, with a build life of a minimum of 120 years. After much deliberation and 10 disparate schemes, Balfour Beatty Construction Company won the project, later to become Transmanche Link Company. Digging on both sides began together, with each construction team aiming to meet in the middle. This was done using sophisticated surveying equipment and lasers. Nobody knew if the two would meet, but on December 1st 1990, the two sides met and two construction workers, Graham Fagg (GB) and Philippe Cozette (Fr) were picked through a competition to be the first to shake hands through the breakthrough hole.

Great Britain and France were joined for the first time in history. On December 10th 1993, the first test run was completed from one end to the other, officially opening on May 6th 1994. After more than £15 billion expenditure and six years of construction, the Channel Tunnel was complete.

Diesel engine Pre-Diesel patent 1890

Herbert Akroyd-Stuart was an inventor and mechanical Engineer noted for his 'Hot Bulb Engine' or heavy oil engine creation. He patented his compression ignition engine two years before German Rudolf Diesel. Akroyd-Stuart built his initial prototypes in 1896 and filed his first patent in 1890. His heavy oil engine was far more advanced than Diesel's engine. Furthermore, it was two years ahead in its submission and patent awards.

Flyball governor 1788

Engineer and inventor James Watt created his first Governor in 1788 after his business associate Matthew Boulton suggested he should pursue the idea and develop it. The Flyball Governor (as Watt named it) was a tapering pendulum based on a version he had developed for his steam engine designs. As the governor's rotation accelerated, fins on top hinges opened out due to the increased centrifugal force being applied and as the fins rose, they were connected to a rotating valve on the central shaft. The valve opened more as speed increased and closed as speed decreased.

Gas turbine 1791

John Barber was an inventor who is recognised as having applied for and been awarded the first patent for a gas turbine design. Barbers creation used a separate reciprocating compressor that forced air through a gas-fired combustion chamber. The hot jet of gas and air mixture was then aimed through nozzles onto a vaned wheel (much like a water wheel). The output of which was sufficient to power both the compressor and its external load. Future generations of jet engineers based their own developments on Barbers ground breaking invention.

Guillotine 1200s

Contrary to common folklore, the guillotine was invented in Halifax, where the lord of the manor had the power to execute anybody he decided had committed such a crime to warrant the beheading of the unfortunate citizen. Often crimes as minor as common theft would suffice a beheading.

Hydraulic crane 1847

William George Armstrong was an engineer and industrialist who invented the high-pressure hydraulic crane. Armstrong also invented a hydraulic accumulator comprising of a large cylinder filled with water and a piston designed to raise the water pressure in the cylinder, then high pressure water flowed down the connecting pipes to the hydraulic pistons on the cranes and other hydraulic machinery. Armstrong's hydraulic accumulator supplied water with high pressure at up to six hundred pounds per square inch. A few years later following a string of successful developments, he was elected a fellow of the Royal Society.

Internal combustion engine 1838

William Barnett was an inventor who applied for a patent for the world's first internal combustion engine design and working prototype. The documents reveal that his submission centred on a double-acting gas engine and detailed a single cylinder, placed vertically with explosions taking place either side of the piston. Barnett's flame method of ignition was very efficient and it was used extensively up to about 1892.

Many inventors attempted to improve on Barrett's design, but none of them overcame the main problem of producing an engine that would be commercially successful.

Jet engine 1930

Sir Frank Whittle was an aeronautical engineer and RAF pilot who filed the world's first patent for a jet engine in 1930 and it was granted in 1932. Whittle was a 24-year-old RAF fighter pilot who first patented a revolutionary type of aircraft engine: the turbojet. It was not built until 1937 due in part to lack of government foresight or backing because its design was so radical a departure from the current aircraft designs. Consequently, he couldn't find enough private backing for the project until 1937, but although he was running his first engine that year, it didn't have its first test flight until 1941. It was a 17-minute debut flight at RAF Cranwell in Lincolnshire in the Gloster Pioneer or G4.

Lift or elevator 1823

British architects Thomas Horner and Decimus Burton designed and installed the world's first public 'lift' or 'elevator' called 'the ascending room' at the Regent's Park Coliseum London, for fee-paying visitors to get a 37-mile panoramic view over the city.

Micrometer 1638

Sir William Gascoigne was an astrologer and scientific instrument maker from Middleton. He invented the micrometer in 1638, designed to quantify very small degrees of measurement, essential in the world of science and precision engineering. Micrometers based on Gascoigne's design are still made and sold to this day.

Multiple-wheel steam turbine 1884

Scientist Sir Charles Algernon Parsons, invented the world's first multiple wheel steam turbine. Parsons first model was attached to a dynamo, generating seven and a half kilowatts of power. His invention allowed the production of cheap and abundant electricity that transformed naval warfare and marine transport.

Razor 1962

Wilkinson Sword made the world's first commercial stainless steel long life blades. Early carbon steel blades either had to be sharpened regularly or disposed of after every shave. Wilkinson Sword's blades, although more expensive, could be used for a week before the need to replace them.

Rubber band 1845

Inventor Stephen Perry patented the rubber band in March 1845. His early products were made of vulcanized rubber. This ubiquitous stationery item is now used the world over, with billions produced and sold each day.

Safety lamp 1815

Sir Humphrey Davy was a scientist and prominent inventor. The first prototype of Davy's safety lamp was developed in 1815, devised to be lit securely and safely for miners to use. Its design prevented the heat from the flame igniting the methane gas found in pits, as the miners bored deeper into the earth.

Screw propeller 1836

Sir Francis P Smith developed the world's first ship's screw propeller. Sir Francis Pettit Smith was an inventor who with John Ericsson designed and developed the screw propeller. Smith was also instrumental in developing the world's first ship, using a screw propeller – the SS Archimedes. He was later key in influencing Isambard Kingdom Brunel to alter his design of the SS Great Britain, from paddle to screw propulsion after lending Brunel the SS Archimedes for a number of weeks for testing and proving.

Screw threads Standardisation 1841

Joseph Whitworth developed and specified the first British national thread standard in 1841. His new standard specified a 55° thread angle, radius of 0.0137329p and thread depth of 0.640327p, p being the pitch. According to Whitworth's chart, the pitch diameter increased in steps. BSW (British Standard Whitworth) is one of three screw thread standards based on the imperial system of measurement, using the same hexagonal nut size and bolt heads, the other two being BSF (British Standard Fine Thread) and BSC (British Standard Cycle). Collectively they are all known as Whitworth threads. The BSW was also adopted in the USA, but was replaced when iron bolts used on railways were replaced with steel. But the USA still used BSW right up to the 1970's for some aluminium parts.

Seat belt circa 19th Century

Sir George Cayley invented the world's first operational seat belt in the early 19th Century. Although Cayley was better known for his huge contribution to the world of aeronautics, he was also responsible for the invention and development of many other significant inventions such as: self-righting lifeboats, caterpillar tracks, prosthetics and optics to name but a few. But he is recognised as being the original inventor who commercialised and applied the use of seat belts, primarily to secure his brave test pilots flying his manned aircraft in the early part of the 19th Century.

Spark plug World's first 1888

Physicist Sir Oliver Lodge was a former principal of the University of Birmingham. After many years of research on electromagnetism, Lodge submitted a paper to the British Science Association in Bath, detailing the phenomenon of electromagnetic sparks in varying gap sizes of two wires when supplied with a current. His early discovery developed into electric spark ignition for the internal combustion engine. His two sons later continued his work and later formed the world's first dedicated spark plug manufacturing company 'The Lodge Plug Company'.

Steam engine First steam engine 1698

Thomas Savery was a military engineer from Devon with a great love of anything mechanical and mathematical. Savery patented an early steam engine described on the submission as 'a means of raising water and occasioning motion to all sorts of mill work by the impellent force of fire…'. Savery subsequently demonstrated his invention to the Royal Society in 1699. His invention had no moving parts such as pistons, just taps. Its operation was very simplistic and unreliable by today's standards, relying on the expansion of the steam to push the water up through pipes.

Steam engine Atmospheric 1705

Engineer and inventor Thomas Newcomen, constructed the first practical machine to employ the power of steam to generate mechanical output. Newcomen engines as they were known, were eventually sold and used throughout Europe primarily to pump water from mines.

Steam engine Modern condensing 1782

Engineer and inventor James Watt, designed and patented the world's first condensing steam engine that produced an uninterrupted rotating motion. Watt's invention, for the first time enabled industrial machines to be powered, being able to be placed almost anywhere so long as a good supply of coal, wood and water was available.

Steam turbine 1884

Sir Charles Parsons developed the world's first steam turbine engine. Following the invention and development of the electric generator by Faraday, the next development was to create a machine that could drive it to generate electricity on a commercial scale. Piston engines were noisy, vibrated a lot and were not reliable, so the steam turbine appeared to be the solution. Today, 75% of the world's power stations still generate electricity using steam turbines based on Parson's original invention.

Telescopic sight 1630

William Gascoigne was an astronomer who was recreating an optical set-up, when a spider's thread became caught between the two lenses. It was positioned exactly between the two lenses by chance, enabling him to see the web thread and the subject both crystal clear. Gascoigne realised that the precision of the telescope was vastly improved if the web was used as a guide. He further refined his discovery by adding another fine thread crossing at right angles enabling him to centre on an object; thus, the telescopic sight was created.

Tension spoked wheel 1808

George Cayley, a prolific aeronautical engineer of his time was looking for a lighter, stronger type of wheel for his new glider to replace the existing solid wood wheels. His improvement of a new design moved the force balance on his wheels from compression to tension by introducing a series of spoke's spanning from a hub to the rim. This was truly a revolutionary invention that fulfilled all of Cayley's wheel requirements, but the idea didn't really develop further until the advent of the bicycle and powered cars. The rest as they say is history.

Thermosiphon Conductive Heating System 1828

Thomas Fowler was an inventor from Devon who developed a device he named the 'Thermosiphon' that still forms the foundation of modern central heating systems. The device was the first conductive heating system, but due to severe patent flaws in the way it protected slight changes outside the main design, many other companies soon copied Fowler's brilliant design.

Thermostat 1830

Andrew Ure, a chemist and medical doctor was passionate about the new steam powered machines and factory systems. This love of technology led him to become one of the most prolific inventors of his time. In the huge textile mills of industrial Britain, a constant temperature was required to produce consistent fabric and other goods. This was a difficult task to achieve and therefore the quality of goods was inconsistent. To combat this issue, Ure patented the thermostat that automatically controlled the temperature to improve the consistent quality and standardisation of textiles into the future.

Tumbler lock 1778

Locksmith, Robert Barron invented the world's first tumbler lock. His patent was submitted in October 1778 and was a critical event in the world of lock making. It was the first lock to feature the 'double acting' principle featuring three tumblers. Prior to Barron's invention, the 'fixed ward' type of lock was the cornerstone of security. A simple bolt was held in a locked or unlocked position by a simple single tumbler catch that located onto the notches at the top edge of the bolt latch. This was a very easy lock to pick and open without a key. Therefore, Baron's three tumbler double acting lock was to lead the way and reset the bar in the world of lock making and high security for years to come.

Tumbler lock High Security 1784

While Robert Barron was fine-tuning his new three-tumbler double acting lock, Joseph Bramah was pursuing a completely different direction in the development of high security locks. In his 'Safety lock' that he patented in 1784, it contained just a small steel tube, with a series of V shaped slots at one end. When the cylindrical key is inserted into the lock, slots accept depressed slides pushed in by the key. The slides must all be precisely the same length otherwise the bolt cannot function. Bramah was so sure that his lock could not be picked that he would give a reward of £200 for anybody who thought they had the ingenuity and knowhow to breach it. His faith in the design was completely justified. It took more than 50 years before a locksmith in the USA claimed the reward

Tyre Pneumatic 1845

Briton, Robert W Thompson was only 23 years old when he filed a patent for the world's first pneumatic tyre. He received grants for them in France (1846) and the USA (1847). The design consisted of a low-profile tube of inflated vulcanized Indian rubber between the rim and tyre, providing an 'air cushion' between the vehicle and the ground. Thompson fitted his new invention to several horse drawn coach wheel sets and demonstrated his 'Aerial Wheels' in March 1847 in Regent's Park, London.

Universal joint commercial design 1676

Robert Hooke was a prolific 17th century inventor. Although not a completely new invention in principle, older designs were not practical or efficient in use and so Hooke set about to develop a truly efficient working design. Hooke's Joint is a coupling or linkage that permits the shaft to bend in either direction, commonly used in drive rods that provide rotary motion and transmission. The device consists of two hinge-like joints in close proximity of each other, set at 90° to each other and linked up with a cross shaft.

Vantablack® World's blackest material 2014

The darkest and blackest material in the world, was invented and developed by a team of British scientists from Nanosystems®, based in Surrey. Made using carbon nanotubes, four times thinner than a human hair, the material is so dark that it is apparently like looking into a black hole. The material absorbs all but 0.035% of light and can be used for a whole variety of applications such as stealth craft, telescope lining and even combat stealth suits for future 'invisible armies'.

Wall plug 1910

John J Rawlings patented the world's first expanding wall plug in 1911. Rawlings was one of the brothers that made up a small plumbing firm in London, Rawling Brothers. Through necessity, Rawling invented the first fibre wall plug, known as the Rawlplug®. Before this point, the main method of fixing items to walls was to hand drill holes, then insert a small wooden wedge into which a screw or nail is inserted. This laborious and crude method of fixing, led Rawlings to experiment with a simple cylindrical dense fibrous tube of varying diameters and these were simply inserted into correctly drilled holes. When a screw was inserted into the middle of this plug, it naturally expanded, preventing the screw to pull out and the rest is history. In 1911 the plug was patented and the following year the trademark was established and registered. In 1919, the company changed its name to the Rawlplug Company Ltd. To this day Rawlplug is still used, as a generic term for any kind of wall plug such was its success.

Wellington boots — Early 19th Century

The World's first rubber boots were made by the North British Rubber Company Ltd, taken from a type of riding boot, Wellesley, named after the 1st Duke of Wellington. They were crafted by hessian backed rubber or leather. The Duke was seen many times in portraits of him proudly wearing these boots, allegedly of his own design. More recently, PVC has been used primarily for its cost, durability and effectiveness.

Workbench Portable — 1968

Mate-Tool, the world's first truly portable workbench was made by ex-Lotus car designer Ron Hickman and later licensed to and made by Black & Decker in 1972 as the now ubiquitous 'Workmate'.

Ethics, Philosophy, Religion, Politics

Amnesty International 1961

Founded by solicitor Peter Benenson, Amnesty International is now represented in more than 150 countries and is the world's largest voluntary Human Rights organisation.

Anglicanism Henry VIII 17th Century

The term Anglicanism is a term used to describe protestant Christianity taught by the Church of England and to Christian faiths that are historically tied to it or with similar beliefs, church structures and worship practices. Although the word Anglican dates back to the Medieval Latin phrase, ecclesia anglicana which translated means English Church, the use of the term Anglicanism was devised predominantly during the reign of Henry VIII; a passionate protestant king, although he still remained a strong believer in the core theological interpretations of the Catholic faith.

Clean air act 1956

Britain was the first country in the world to introduce a 'clean air' law. It was introduced to combat the smogs of the first half of the 20th century, which in turn caused many deaths through pollution. It bans emission of smoke within designated areas, predominantly densely populated towns and cities. The act prohibits emissions of dark smoke from any chimney, whether industrial, trade or private premises and bans cable burning except under license. This resulted in the majority of towns and cities in Great Britain, having some of the cleanest urban air quality in the world. Unfortunately, vehicle ownership was growing fast and the country was then faced with another pollutant to control.

Free trade theory 1846

Dr Andrew Ure, a British chemist, became a pivotal advocate for the advantages of free trade unregulated domestic production. Ure published two books in the 1830s: The Philosophy of manufactures (1835) and Cotton manufacture of Great Britain (1836), in which he argued that by revoking protectionism, England would become immensely wealthier. This he suggested would benefit all the people; while at the same time improve European and US relations. Conversely, Ure argued that protectionism created hostility that often developed into War. Ure backed up his arguments with an notable selection of data and pictorial charts to stress the correlation between the wealth, productivity and wellbeing of nations.

Habeas Corpus 1215

The literal translation of this Latin term is 'you may have the body'. It is effectively a court order or writ that requires any person who is arrested, to be taken to court before a judge. Habeas Corpus requires that an arrested person can be released from what is considered unlawful confinement, due to lack of evidence or cause. This right originated in the English Legal procedure in the 13th century and is now one of the principle foundations of many other legal systems around the world. Before the signing of the Magna

Carta in 1215, the ruler or his men could confine a person to prison for absolutely any reason. It now forms an integral thread in the cultural DNA of Great Britain.

Holiday pay for workers 1872

George Livesey was a British industrialist and philanthropist and owner of the South Metropolitan Gas Company. Livesey provided the world's first paid holidays for all his manual workers in 1872.

Jury trial 1166

Trial by jury came about during the reign of Henry II. The basis of which has not changed for almost 1000 years and still forms the foundation of trial by jury in most democratic countries throughout the world.

Keynesianism 1936

John Maynard Keynes was a British economist, born 1883 in Cambridge, who studied at Eton College and as an undergraduate at King's College Cambridge. His areas of expertise were macroeconomics, liquidity preference, deficit spending and AD-AS modelling. Keynes is best known for his own version of macroeconomics, now universally known as Keynesian economics and in particular his work on business cycles. He is now widely recognised as one of the founders of modern economics.

Liberalism Classic 1632 to 1704

John Locke was a physician and philosopher, born in Wrington, Somerset in 1632 and known as the Father of Classic Liberalism. Locke was widely regarded as one of the first of the British empiricists and a pioneer in the philosophy and theory of social contract. Locke's work had a significant impact on the future development of political philosophy in general and particularly the subject of epistemology.

Magna Carta World's first bill of rights 1215

King John signed the Magna Carta, Latin for great charter, on June 19th 1215 at Runnymede, on the banks of the River Thames in Surrey, just 20 miles west of central London. This was the world's first bill of rights for all its citizens. The Magna Carta was the first ever, formal agreement between a ruler and his or her people, that ensured their rights against the wishes of kings and queens to come, ensuring that any future ruler had to obey the laws of the land. What this ground-breaking agreement did was to ensure that no person could be arrested, jailed or lose their possessions to the ruler or state, unless judgment has been agreed and complied by the law of the land or their equals. Consequently, the Magna Carta paved the way for trial by jury, meaning for the first time in the world, accused people would be tried by their peers, ensuring the civil rights of the person. Thus, for the first time ever, the Magna Carta lay down the principle that all the people of England, represented by the Barons of the country at this period, could constrain the dominance of a king or other ruler if he was carrying out actions or deeds not considered in the country's best interest.

Methodism John and Charles Wesley 18th Century

The movement of Methodism is based on protestant Christianity and is derived from the teachings of British born brothers, John and Charles Wesley. Methodism formed its roots when the Wesley brothers created a movement within the Anglican Church of England in the eighteenth century. The catalyst to Methodism was the Holy Club, formed by the brothers while they were undergraduates studying at Lincoln College, Oxford University. The Holy Club met each week and its aim was to set about creating a set of guidelines and rules for its members to live a wholesome and holy life.

The followers later became known as Methodists by outsiders, due to the manner in which rules and methods were used to live their religious lives. Following enthusiastic missionary pursuit overseas, Methodism grew throughout the British Empire and played a huge part in what is known as the First Great Awakening in colonial America. The now famous Armenian religion in the USA is a development and offshoot of the original Methodist protocol of religious teachings.

Modern economics 1776

Adam Smith was a moral philosopher and economist who wrote the pioneering publication known as The Wealth of Nations. His work is one of the first transcripts describing collectively and probing what develops and builds a nation's wealth. This book is now regarded as the most fundamental work in classical economics that addresses subjects such as productivity, free market dynamics and division of labour.

Modern sociology 1767

Adam Ferguson was an historian and philosopher born in Logierait, Perthshire in 1723 and widely known as 'the father of modern sociology'. Ferguson criticised modern commercial society because he thought it made men weak, greedy, unconcerned for their neighbours and dishonourable. Instead, Ferguson's sympathies and teachings lay with the more traditional communities, who he argued produced loyalty, courage and empathy with their fellow neighbours. His teachings and beliefs formed the basis for most modern social thinkers and socialist political parties in modern times.

Parliamentary democracy 1236

The very first documented use of the term was in 1236 when King Henry III suspended a law case to 'parliament', due to meet in January of that year. This was written in an official document belonging to the English crown. However, this was not the birth of parliamentary democracy. The term used in 1236 merely described an assembly that had existed for countless centuries previous. The word parliament originally evolved from the French word parlement meaning effectively 'discussion'. The assembly in 1240 started to develop features that distanced it from the meetings of old.

The first significant change was the granting right allowing Parliament to set and raise taxes to give to the king. Henry III was the first king to ask for regular tax payments to help fund his military exploits. Most of the future parliamentary meetings were held in Westminster.

However, around 1320, the role of setting and collecting taxes fell squarely on the members of parliament or MPs. Following warfare between England and France and then Scotland, the commons established itself firmly in parliament alongside the Lords and Noblemen's House. However, MP's described the unbalanced relationship between the Lords and the Commons clearly in 1399 when they stated that they were purely 'suitors and petitioners', whereas all 'parliamentary judgments' belong exclusively to the lords and kings. This was clearly the first stage on the road to parliamentary democracy, as we know it.

Quakers George Fox 15th Century

The Quakers is a religious society of friends founded by English dissenter and philosopher, George Fox. Fox was born in Drayton-in-the-Clay, Leicestershire in 1624. During his adult life, Fox rebelled against the political and religious authorities by proposing a completely new advance in religious thinking to the Christian Faith. His ministry eventually grew to North America, but he spent his final days living and working in London establishing the ever-growing Quaker movement. Unlike most other Christian groups, the Quakers have actively avoided hierarchical structures and creeds. Today, there are more than three hundred and sixty thousand adult members worldwide meeting on a regular basis. Many famous Quakers have mixed business with religion and formed such huge institutions as Barclays and Lloyds Banks, Friends Provident, Clarks Shoes, Cadbury, Rowntree and Fry's chocolate companies.

Salvation Army 1885

William Booth came from Nottingham, one of five children born to Samuel and Mary Booth. In 1849, he moved to London searching for work and in the process joined a chapel in Clapham, where in later years he met his future wife Catherine Mumford. He became an evangelist in the Methodist New Connexion. After different posts within the church he resigned in 1861 due to 'too many restrictions' on his faith and beliefs. He formed the Christian Mission in 1865 and Booth was its leader.

In 1878 the Christian mission was renamed The Salvation Army. He gave himself the grand title of 'General Booth' and described this new organisation in the following way: "we are a salvation people – this is our specialty, getting saved and keeping saved, then getting somebody else saved". There was often disagreement to the 'Army's' ways and methods in the early years. After his wife died of cancer in October 1890, this left huge emptiness in his life. In that same month, he published a major social policy called 'in darkest England, the way out'. He investigated many ground breaking notions including, employment centres, helping young men learn a trade and get involved in agriculture before they emigrated and ideas such as providing hostels for the homeless.

Following a difficult period with three of his children leaving the Salvation Army and one dying in a train crash, he decided to hand over the reins to his eldest son, Bramwell. In 1904, William Booth instigated his first ever motor tour, travelling from Land's End in Cornwall to Aberdeen in Scotland. He carried out a further six tours in Great Britain. Then in 1905, on the way to Australasia, Booth stopped off and visited the Holy Land to take in many of the biblical sites. When he returned to Great Britain, he was given the freedom of the Cities of Nottingham and London and he was further awarded an honorary doctorate in Civil law from the University of Oxford.

Despite failing eyesight, he still carried on with his campaigns culminating in a trip to Norway in 1912. It was in this year of 1912 that the General finally met his saviour. He left behind a legacy that was the Salvation Army operating in 58 countries and run by a faithful army of more than 15,900 'soldiers'.

Utilitarianism Jeremy Bentham 16th Century

Jeremy Bentham was a jurist, social reformer and philosopher, born in London 1748. Bentham is recognised as the founder of modern utilitarianism, the theory that suggests that the correct course of action is the one that maximises total benefit and reduces suffering to the bare minimum for all. It is an ethical human centric theory for economics rather than a market centric philosophy. Utilitarianism centres on the importance of happiness in any economic outcome.

Women's liberation 1901

The First World War gave women the first opportunity to challenge jobs that were traditionally for men and it was in 1918 that women had the right to vote if they were over 30, giving them the same political rights as men. Getting the vote was only part of the improvements in the rights for women.

The women's rights movement had also campaigned for the law to be changed so that a woman could divorce her husband, the right to education, the right to go to University and the right to become a doctor. But with the advent of the suffragette movement, women were campaigning for voting rights on equal terms with the men.

The hero and legend of the suffragette movement was Emily Davison, who felt so passionately about her cause that on 13th June 1913, she ended up running in front of the King's horse at the famous Epsom Derby. However, rather than stop the race to make a point and generate huge publicity for her cause, she suffered horrific injuries and died four days later in hospital. There was great public condemnation for her actions. However, gradually, the fortunes and aims of the women's movement meant that although tragic, Emily Davison's death was not all in vain.

> "Ethics is knowing the difference between what you have a right to do and what is right to do"
>
> Peter Stewart

Fabric production

Fabric manufacturing and process

Dye Synthetic 1856

William H Perkin was born in East London in 1838. He was a British chemist, who was admitted to the Royal College of chemistry at the tender age of 15 and was most famous for his unintentional discovery of synthetic dye at the age of 18. Perkin's discovery started during trials at the college, after he commenced on a series of experiments to try and manufacture synthetic quinine used to treat malaria, as a substitute due to its high cost.

It was during experiments in a make shift laboratory at the top floor of his home in London, that Perkins made a great-unplanned breakthrough: that Aniline could be transformed into a rather basic mixture, that when extracted with alcohol produced a substance of a deep purple colour. He immediately became enthusiastic and with the help of friends decided to continue experiments to scale up production of the purple liquid and commercialise it as a stable colour fast dye, that they named Mauveine. Perkins applied for a patent in August of 1856 while still only 18. After an illustrious working career, in 1906 Perkin was Knighted, fifty years after his initial discovery of Mauveine. It is hard to imagine a world without multi-coloured fabrics, but we now take this marvel for granted thanks to Perkin's great discovery.

Elastic fabric 1830

Inventor Thomas Hancock developed fabric elastic. Hancock reformed naturally stretchy rubber in a machine that mixed leftovers of rubber, crafting it in such a way through high-pressure rollers and forming sheets that increased its elasticity. He later developed this process for mass commercial use.

Fibre Man made 1883

Inventor Joseph Swann produced the world's first synthetic or man-made fibre in 1813. Swann extruded a mixture of cellulose nitrate through a grid of small holes. Rayon fibres spun from what is now known as Viscose later replaced these early man-made fibres. Today, the majority of fabrics have a percentage of man-made fibres woven into them to strengthen against wear.

Polyester 1941

John Whinfield and James Dickson were scientists, who together invented polyester in 1941 and because of its flexibility in making artificial fabrics it became an overnight success. Polyester is made from oil, or to be more precise, polyethylene terephathalate that originates from petroleum. Its invention was declared a wartime secret due to its many uses during severe raw material shortages in WWII.

Knitting machine 1589

Reverend William Lee was a clergyman and inventor who produced the world's first knitting machine, used in its original design for centuries after its creation; its key functional concept remains the same today. Lee was said to have created the knitting machine after noting that the woman he was courting at the time, devoted more time to knitting than him. The first version of his machine in 1589 produced stockings using coarse wool, but when he applied for a patent Queen Elizabeth I refused it. So he developed a second, more refined knitting machine that produced a much finer silkier texture, but his patent application was again denied by the Queen, said to be due to her concern for the job security of thousands of hand knitters throughout her Kingdom.

Disheartened, Lee travelled to Rouen in France with his brother and with the support of Henry IV of France, he commenced stocking production and rapidly grew his business with huge commercial success up to Henry's assassination in 1610. Following Lee's death, a year later his brother returned to Britain and gradually established his fledgling semi-automated knitting industry, often fighting fierce opposition from the hand knitting community.

Loom Powered 1785

Inventor and Industrialist Edmund Cartwright, invented the first Power loom that combined threads to make cloth. It was driven by steam, in a factory that Cartwright had set up in Doncaster for cloth manufacture. Being the first, it could always be improved and over the next few years, Cartwright refined his process, benefitting from higher production output and lower maintenance downtime.

Mackintosh Waterproof Coat 1823

Charles Mackintosh was a chemist and inventor of waterproof material that led to the creation of the Mackintosh raincoat. It was through experimentation with the by-products of tar that led Mackintosh to discover a waterproof covering for fabrics and eventual patent application for his process. Essentially, Mackintosh's invention was based on glueing two fabrics together with Indian rubber, made soluble by Naphtha. For many years to come, the term 'Mackintosh' or 'Mac' became the household name for a raincoat.

Mercerized cotton 1850

John Mercer was an inventor from Great Harwood. Mercer's treatment of cotton fabric using sodium hydroxide to swell the yarn and give it a lustrous sheen is now known as 'Mercerizing'. Mercer's process is still used internationally in the production of fabric.

Rayon Viscose 1892

Of all man-made fibres in the world today, Rayon is the oldest of them all. It was the first viable manufactured fibre. English Chemist Charles F. Cross, and his collaborators Edward John Bevan and Clayton Beadle, discovered the process of making Viscose in 1892.

Rayon Acetate 1895

The cellulose acetate manufacturing process was invented and produced initially by Charles Frederick Cross and Edward John Bevan (of Viscose fame). It was used by the pair as an alternative to their original invention of Viscose Rayon, due to its differing texture.

Sewing machine 1790

Thomas Saint, the inventor, is generally recognised for developing and making the world's first sewing machine design, but like so many inventors, did not adequately market or sell his invention. His machine was designed to be used on canvas or leather material. The device itself featured a feed mechanism, overhanging arm, a looper and vertical needle bar, using the chain stitch method where the machine used a single thread to make simple stitches on the leather or canvas. Saint's Sewing machine was very advanced for its time, but its undeveloped and untested concept meant that it would need much refinement and development over the ensuing years to become a truly practical proposition.

Spinning Jenny 1764

inventor, carpenter and weaver James Hargreaves, developed an improvement on the established spinning wheel by creating the 'Spinning Jenny'. Hargreaves' device was a hand turned spinning machine that powered eight spindles to create an eight-threaded weave. Advanced models later in its development used up to one hundred and twenty spindles. Like so many technological advancements, Arkwright's machines were a threat to traditional fabric workers, so much so that in 1768 a group of spinners destroyed Hargreaves' Spinning Jenny machines after breaking into his house, fearing their jobs would be lost forever.

Spinning frame 1768

The Spinning frame was invented and developed by industrialist Richard Arkwright and it became the keystone of the global textile boom at the start of the industrial revolution. Arkwright's spinning frame spun thread that had a much tighter weave than previous incarnations and because of the new fabric density, was far stronger. Britain became known as the 'Workshop of the world' through this single key invention, transforming the textile industry in England's North West.

Spinning mule 1779

Industrialist and inventor Samuel Crompton, created the spinning mule in 1779. Crompton's invention combined the moving frame of the spinning Jenny with the water frames rollers. His device gave the operator greater control over the weaving process, enabling many different sizes and texture of yarns to be used.

> 'The UK is the 15th largest textile manufacturer in the world and exported £6.5 billion of clothes in 2017"
>
> Financial Times

Food and drink

Champagne or sparkling wine 1662

Christopher Merrett was a physician and scientist from Winchcombe in Gloucestershire. He was the first person to document the intentional addition of sugar for the manufacture of sparkling wine. In his eight-page paper, Merrett refers to a secondary fermentation process, a chemical reaction occurring when the bottled alcohol underwent increases in temperature, thus producing carbon dioxide. This process now forms the basis for all Champagne production. The French have always maintained that Sparkling Wine or Champagne (as it is now known), was invented by the Monk Dom Perignon in 1697 by accident. But Interestingly, the earliest French documentation is 1729, an incredible 67 years after Merrett's invention.

Champagne bottle design 1635

Admiral Robert Mansell discovered that adding iron and magnesium to the glass making process, made a stronger glass to contain high gas pressure. Sparkling Wine inventor and Briton, Christopher Merrett, needed a bottle that would not explode under pressure from his new sparkling wine and later adopted Mansell's design. Contrary to common belief this was not a French design and was created long before French sparkling wine started being produced. Mansell's original bottle design also predated Merrett's sparkling wine.

Chocolate bar 1847

Joseph Fry made the world's first chocolate bar, developed and created by his company JS Fry & Sons of Bristol in 1847. Up until this point, chocolate had only been available as a drink. The secret was to mix cocoa powder with cocoa butter and sugar, ensuring the product remained solid at room temperature, but melting in the mouth when eaten. It was billed originally as 'chocolate delicieux a manger', or delicious to eat.

Chocolate bar World's most popular 1886

Made by British confectioner Rowntree's, Kit Kat® is the world's most popular chocolate bar or snack by numbers sold. 17 Billion Kit Kat® bars are sold each year in more than 70 countries. Far more sales than for any other confectionary item globally.

Chocolate Drinking chocolate 1689

Physician Hans Sloane introduced milk to chocolate for drinking, while working as a Doctor in the West Indies. He found it very hard to digest, so he decided to boil the beans and add milk and sugar to make it more palatable. As soon as Sloane arrived back into Great Britain 'Sloane's Milk Chocolate' was being sold all over London. The recipe was eventually passed on to the chocolate giant Cadbury's in Bournville, Birmingham.

Coffee houses Coffee shops 1650

The real coffee shop revolution didn't start in recent times with the large multi nationals. In fact, it first started in 1650 in Oxford and following its success moved to London in the Cornhill district, taking this new drink to the gentry of this bustling city. It also became the birthplace of the London Stock Exchange. Prior to coffee shops, traders had carried out their business in taverns that were by their nature, raucous and unpleasant. By comparison, Coffee prevented drowsiness and sharpened the mind, spawning the beginnings of what is now known as the stock market.

Cornish pasty 1509

With a history dating back to the original tin industry of Cornwall, the Cornish pasty was developed and made by wives to feed their hungry tin mining husbands and sons. Affectionately known as an 'oggie' by locals, the traditional Cornish pasty consists of beef skirt, potato, onion and turnip, wrapped in a case of pastry and deep crimped around the edges. The crimped crust served as a handle to hold the hot pasty when down in the mine and the thick short crust pastry acted as a superb insulator, keeping the hand-held meal hot before eating. After the demise of the Tin Mining industry, many experienced miners emigrated and took their recipes with them. You can now get a great pasty in Argentina, Australia, New Zealand, Brazil and many other South American countries.

Fish & chips 1860

For many years, Fish & Chips have been the national dish of Great Britain, often symbolising its great cultural and culinary heritage and like the Union Flag, recognised as British the world over. Fish and Chips were first served together as a single dish around 1860 with both Manchester and London laying claim to being first. But fried fish and fried potatoes had been in existence in Britain for hundreds of years prior to the joining together of this great gastronomic marriage.

Fizzy drinks Carbonated drinks 1772

Eighteenth century Scientist and clergyman Joseph Priestly, was the inventor of carbonated water. He did this by suspending a tub of water above a large fermenting ale container at a local brewery in Leeds and the water above rapidly developed a pleasurable gaseous syrupy and acidic taste. He published a description of how to produce his soda water in 1772. Several years later, Johann Schweppe created Schweppes and started manufacturing fizzy drinks using the Priestly method.

Full English breakfast circa. 1300

The traditional Full English Breakfast has been a staple part of the British diet for many years with many variants such as Welsh, Scots and Irish Breakfast. The main ingredients remain similar, being bacon, pork sausage, egg, baked beans and mushrooms, with a side of toast, a cup of hot tea and optional black pudding.

Gin & Tonic circa. 1820

Widely regarded as the world's most civilised cocktail, G&T was created by British soldiers based in India. Tonic water contained Quinine a malaria preventative that had a bitter after taste and the gin was used to sweeten the drink.

Haggis circa 100AD

Haggis is now a Scottish traditional savoury food made from minced sheep's heart, liver and lungs, mixed with onion, spices, suet and oatmeal all mixed together with stock. The ingredients are traditionally cooked in the animal's stomach lining. There is evidence that the haggis dates back as far as Roman times, to circa 100AD.

Sandwich 1762

John Montague, the fourth Earl of Sandwich, coined the word 'sandwich' according to historic legend. It is doubtful that he actually invented what we now know as a sandwich, but he certainly made it popular and gave rise to its name. It was apparently in 1792 when he was said to have asked for meat to be served between two slices of bread, in order to avoid interrupting a gambling game. This story may have been put about to create adverse publicity by his rivals, but the term Sandwich stuck and is now a household name used worldwide. The Earl had no actual connection with the English coastal town of Sandwich, only by title. Proving that British hereditary titles can be very confusing. The Sandwich Isles were also named after the fourth Earl of Sandwich.

Soft whipped ice cream 1949

Prior to a career in politics, ex-Prime Minister Margaret Thatcher took on a junior research role as part of a development team at food manufacturer J. Lyons and Co, putting her Oxford chemistry degree to good use. The main focus of her role was to develop a method to introduce more air into ice cream to produce a smoother whipped texture that could be forced through a tube, forming part of an ice cream delivery system. This system proved popular and was the catalyst for '99 ice creams' and brands such as 'Mr Whippy''.

Sunday roast circa.1480

The archetypal Sunday family meal for centuries in Britain; traditionally, roast beef was the main meat and is still the national culinary custom. But since WWII, lamb, chicken and pork are just as popular now on the British dinner table. It is also called a "Sunday joint," and is traditionally served midday on Sundays with roasted potatoes, Yorkshire pudding, vegetables and gravy with an optional side of horseradish if serving beef. Lamb is served with mint sauce; pork with apple sauce and chicken or turkey is served with light poultry gravy, with redcurrant or bread sauce.

Tin can 1810

Up to 1810, food could only be preserved by either salting heavily or storing for limited time in glass jars, by cooking it for hours first to help sterilize it. However, British Merchant Peter Durand took the jar preservation method a step further and decided to use a tin can. The main problem was that either a sharp thick blade or a hammer and chisel were required to open the cans. It would be another 48 years before American Ezra Warner invented the first tin can opener!

Welsh rarebit 18th Century

Welsh Rarebit is a traditional snack established in the eighteenth century, originally from Wales. It is essentially grated cheese melted with milk or ale and mixed with butter, mustard, pepper and salt. This mix is then spread to cover pre-prepared toast and grilled or baked until the cheese bubbles and browns.

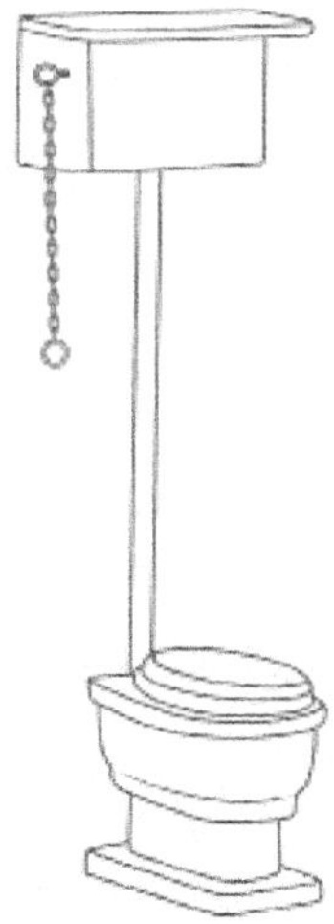

Flushing toilet

Household and general innovations

Anywayup cup 1990

Inventor Mandy Haberman designed the world's first no spill toddler training cup in 1996. After further development, the cup became an instant success and is now sold to millions of parents across the globe.

Baby buggies Strollers 1965

Aeronautical designer and test pilot Owen MacLaren, produced the world's first lightweight aluminium pushchair that folded using one hand, developed from his expertise in designing the folding undercarriage for wartime Spitfires. Following testing, he then started manufacturing them in 1967 and now the name MacLaren is synonymous with the folding baby buggy.

Bagless vacuum Cleaner 1983

James Dyson is a British inventor born in Cromer, Norfolk in 1947, most recognised for the creation of the bagless Dual Cyclone vacuum cleaner. It works via cyclonic separation and unlike conventional vacuum cleaners, does not lose suction as it fills up with dirt. The idea originated as a development of both the centrifugal oil cleaner used in oil refineries worldwide and from his spray finishing room's air filter in his ball barrow factory (an earlier Dyson invention).

Dyson launched the G-Force cleaner in 1983, although at the time no manufacturer wanted to take on a new cleaner without the continued after sales requirement of dust bags. This prompted Dyson to launch it instead in Japan through local catalogue sales. He obtained his first US patent in 1986 for the principle and later won the 1991 International design fair prize in Japan. Dyson then set up his own manufacturing company in the summer of 1993 with a research centre combined in Malmesbury, Wiltshire. Dyson's subsequent TV campaign focused on users not having to continually buy dust bags, with the slogan 'say goodbye to the bag', proving more successful with the public than his previous focus on suction.

The Dyson Dual Cyclone became the fastest selling vacuum cleaner made in the UK, outselling some of the companies that had initially rejected his idea and is now one of the most recognised and innovative brands in the UK. Dyson cleaners had become the market leader in the USA during 2005 by value, rather than units sold.

Ball barrow 1974

In 1974, James Dyson, the prolific inventor, designed the ball barrow. It featured a steel frame supporting a plastic moulded hopper and a front mounted spherical plastic wheel. It was developed through the restrictions of conventional wheelbarrows using thinner pneumatic front wheels, that were less controllable and prone to getting punctures. The ball wheel also gave the barrow stability in soft ground surfaces making it more stable laterally under load on more uneven ground. The Ball Barrow was awarded the British Design Innovation Award in 1977.

Banknotes First public issue 1797

The Bank of England issued the world's first commercial banknotes in 1797 after the small-scale Napoleonic invasion of Fishguard in Wales, caused jitters amongst its hierarchy and forced the bank to issue IOU bank notes. The plan was for Napoleons second rate army to land near Bristol, destroy it and then move on into Wales, marching North towards Liverpool and Manchester. It is recorded that the Bank of England took fright at the Fishguard Invasion because although unsuccessful, they feared a similar more successful invasion might happen again. The bank then decided not to use the existing gold and silver coinage anymore and instead printed banknotes on paper in one and two pound denominations.

The banknotes were effectively IOU's with the declaration 'I promise to pay the bearer on demand the sum of…' still printed on British banknotes. The paper notes were printed on special watermarked paper to prevent copying. From this day on, the world's first bank notes appeared in general circulation.

Bathtub Cast iron enamel bath 1870

Enamel cast iron bathtubs were developed and invented by Briton David Buick. He moved from Great Britain to Detroit Michigan with his parents as a young man and worked in the sanitary and plumbing industry. He developed and invented the internationally successful process for bonding porcelain enamel to cast-iron bathtubs. In 1903, pursuing his interest and love of motorcars, Buick founded the Buick Motor Company. The rest is history.

Corkscrew 1795

Samuel Henshall patented the world's first corkscrew. He was an English clergyman and it is thought that he designed the corkscrew through his frustration of releasing corks from church wine bottles, which until this point was very haphazard and often dangerous to the 'de-corker'.

Electric cooking utensil 1874

St George Lane-Fox, a prolific inventor and electrical engineer took out a patent for an 'electric cooking utensil' in 1874. There is little information on how the patent was commercially exploited, but it is clear that this was the very first example of electric cooking, as we know it today.

Eraser 1770

Edward Nairne was an optician from Sandwich. In 1770, one of the first references to the word 'rubber' was discussed, when Nairne sold natural rubber cubes at his shop in Cornhill. Nairn's cubes were erasers, sold for the extraordinary sum of £3 per cube. Edward Nairne is widely credited with creating the first rubber eraser.

Friction match 1827

John Walker was a chemist who invented the world's first friction match. Like so many great inventions the friction match principle was discovered purely by accident, when a wooden splint encrusted with chemicals burst into flame when scratched across Walker's fireplace hearth at his home. He further developed his discovery and made the first friction match in the world. Walker's friction matches transformed the portability and application of fire. Against the advice of friends, Walker failed to patent his invention and consequently in 1829, Samuel Jones from London copied his method and launched 'Lucifer's', a blatant copy of Walker's 'friction lights'.

Fridge Principle of heat exchange 1748

William Cullen was a physicist and chemist from Hamilton. Cullen demonstrated the first known artificial refrigeration process at the University of Glasgow in 1748, but did not use his discovery for any commercial or practical purpose. To this day, Cullen's principles of heat exchange, employing cooling by rapid expansion of gases, are the main source of refrigeration in all modern systems.

Kettle 1891

The electric kettle was invented to satisfy the huge demand and need of British tea drinkers during the late 19th Century and this invention was credited to Compton & Co in 1891. Several improvements were made over the years, including the British Birmingham based Swan Company who in the 1920s developed the world's first kettle with a built-in heating element.

Kettle Automatic electric 1955

Founders Bill Russell and Peter Hobbs of Russell Hobbs Co Ltd altered the course of standard electric kettles in the 20th Century by designing and introducing their revolutionary vapour controlled K1 automatic electric kettle design in 1955. This British made and designed K1 was the world's first fully automatic kettle. When it was first launched, demand far outstripped production such was its popularity. Before the K1 was launched, old style electric kettles suffered from boiling dry when unattended, creating fire hazards and were a danger to the home, not to mention the element burning out if left without water cover and needing to be replaced each time this happened. This ground-breaking design was so effective and successful that it is still used the world over in all electric kettles sold today.

Lawnmower 1827

Inventor and landscaper Edwin Beard Budding developed the world's first mechanical lawnmower (19 inch) and because of its initial high price to make and buy, it remained the preserve of the very rich until such time that mass production offered lower cost alternatives.

Lawnmower Steam Powered 1893

James Sumner, an inventor from Leyland in Lancashire, patented the world's first lawnmower powered by steam. Known as the Leyland

Steam Lawnmower, it was operated by just one skilled Engineer, who was required to be an expert in steam power, its maintenance and operation. The boiler had to be constantly topped up with water and the machine needed regular meticulous maintenance like any steam driven machine, making running costs very high. In 1895 after moderate success in sales, Sumner renamed his company the Leyland Steam Motor Company a predecessor of the infamous British Leyland Motor Company of 1970's fame. It was not until the introduction of widely available petrol powered internal combustion engines, that powered lawnmowers became a large market commodity.

Lava lamp 1963

Edward Craven Walker was an accountant and inventor who originally created the design after seeing an egg timer in a pub. He set about developing his design based on the egg timer, experimenting with different liquids, until at last he launched his first commercial Lava Lamp.

Meccano 1898

Frank Hornby was an inventor who decided to create a system of parts including nuts and bolts, spanners and screwdrivers, to keep his two young sons occupied through the long winter days. Hornby wanted his boys to be able to recreate the huge cranes that enthralled them so much at their home in the port of Liverpool. Hornby set about developing and commercialising his kit of parts and started to market and sell them under the brand name 'Mechanics Made Easy' in 1901. The name 'Meccano' was patented in 1907.

Hornby's first Meccano factory opened its doors in Liverpool the same year, but as demand was so huge, a much larger plant quickly replaced it. The popularity of Meccano soon spread across the globe and at one point in 1951 just one of Meccano's many plants, the Bobigny factory produced more than half a million boxes in a day. Meccano is still sold in huge quantities the world over; such is its continued demand.

Mousetrap 1899

Born in Leeds, James Henry Atkinson was the inventor who created and patented the spring-operated mousetrap called 'the little nipper'. His design is still sold to this day and many regard the design as being the most effective means of catching mice.

Pram Perambulator or baby Carriage 1733

William Kent designed the first portable pram or baby stroller. Kent was asked to build a device that would transport his children and he duly designed and built a three-wheeled frame with a basket and handles. The baby or infant sat in the basket. Kent's design is remarkably similar to the three-wheeled prams available today.

Safety Pin 1849

Inventor Charles Rowley was regarded as the creator of the safety pin, similar to those in use today. A sliding clasp covered the point of the pin to protect the user with a simple spring joint at the other end to aid opening and closing of the device. Walter Hunt (USA) also made a similar device at around the same time but sold his design to a US company who ended up making millions of dollars in profit over the years to come.

Shaver Electric 1913

Inventor G.P. Appleyard patented a 'power driven shaving appliance', although the design was revolutionary, the device could not be manufactured because motors could not be made small enough at the time.

Shoelaces 1790

Harvey Kennedy is credited as being the inventor of the modern shoelace. He also patented the 'shoe fastening lace' in 1790 and records suggest that he profited a great deal from the proceeds.

Steel pen 1780

It is documented that Samuel Harrison, a manufacturer of split pins, made a steel pen for Dr Joseph Priestly in 1780 and is the first recorded use of steel as an ink pen rather than a feather quill.

Stove Gas 1826

With the advent of coal gas producing companies being firmly established in most major towns and cities in Britain, James Sharp spotted an opportunity to replace the existing coal and wood fired ovens with something altogether more practical. Having constant fuel on tap, the gas stove was developed and patented by British inventor James Clegg. The gas stove was a more efficient method to cook food and it is testament to James Sharp's original design, that gas stoves are still being used today and are often still the first choice for professional chefs the world over.

Tea maker Automatic 1902

Albert E. Richardson was a clockmaker who invented and patented the world's first automatic tea-maker. Methylated spirits were lit by the automated striking of a match, thus heating the hot water and this action then started the chime of a large portable alarm clock bell when the tea was ready. The Automatic Water Boiler Co Ltd was contracted to make the machine.

Toaster Electric 1893

Inventors, Alan MacMasters and Rookes Evelyn Bell Crompton's main specialities were the fields of Municipal Electricity Supplies and lighting and they founded the company Crompton & Co. British inventor Alan MacMasters approached Crompton with an example of a device for heating bread by the use of an electrical element. Masters and Crompton worked together to perfect the design and it later went into production that year. This was the world's first electric toaster and it was named 'The Eclipse'.

Toilet Flushing 1596

John Harrington installed the first non-gravity flushing toilet at his own home in Somerset. Harrington made one for Queen Elizabeth I (his God Mother) and one for himself. He only ever produced these two and was often mocked for his invention, so much so that it was not until 1775 that Alexander Cummings made huge progress in the design of his water closet and momentum then gathered.

Toilet S-trap 1775

British inventor Alexander Cummings filed the world's first patent for an 'S-Trap'. The flush toilet was emerging in its modern form as a device. But it was Cummings' invention of the 'S Trap' that really advanced toilets in the mid 18th Century and is still in use worldwide today. This simple device uses the standing water in the bottom 'U' section of the S trap to act as a seal for the bowl outlet to stop the escape of foul air entering the room from the sewer.

In the mid 19th century, it also prevented vermin and all manner of wildlife from entering the room from the toilet bowl. As an added precaution, Cummings design also incorporated a sliding valve just above the valve in the bowl that remained shut until the toilet was flushed.

Toilet Public 1851

John Neville Maskelyne created the world's first public toilet using Water Closets or WCs, and opened his first establishment at 95 Fleet Street, London. Maskelyne from Cheltenham, had invented the world's first pay toilet. The WCs were initially installed temporarily to help cope with the millions of people that would be descending on London to visit the Great Exhibition. There was a small charge of one penny to use the toilets; hence when needing to use the toilet, the term 'to spend a penny' was commonly used. The toilets became a permanent fixture and then started being deployed, not just in London, but many other towns and cities across the country.

Toothbrush c1770

William Addis designed and made the world's first toothbrush while in Newgate prison after being caught up in riots and then arrested. The businessman used an old bone and the hair of a pig to make the first toothbrush. Necessity is the mother of invention and following release from prison, he built a global powerhouse and the brand still exists today, known as Wisdom Toothbrushes.

Umbrella 1868

Samuel Fox was an industrialist and is best known for his invention of the Paragon steel umbrella frame. His U section steel frame was far superior to his competitors at the time, who all made their frames from whalebone. His design was so successful that production of Fox frame umbrellas only ceased production in Britain in 2003 and the company ceased trading in 2008.

Vacuum cleaner 1901

Herbert Cecil Booth was observing a railway coach in the process of being cleaned whilst out of service. The cleaner used a machine that just blew the dust away and in the event, ended up making more mess inside the coach than was intended. This prompted him to create a device that would work in reverse and suck up the dust and deposit it into a container, thus leaving the coach dust free.

While developing this new machine, he tested his theory by placing dusters or handkerchiefs on the floor and then observed that dust was sucked up and stuck to the cloth on both sides. This was the breakthrough that he was looking for, because it prevented dust flowing through to the motor and causing all manner of problems. Once he fully developed a filtered version, he then set about starting a company using hoses from a purpose converted van on the street and taking the hoses through doors and windows to clean houses. This was the birth of the first Vacuum cleaner, as we know it.

Vacuum flask 1892

Sir James Dewar, a Cambridge based professor and prominent figure of the Royal Institution, invented this simple device. He didn't create it to keep his tea or coffee hot, but to act as an experimental aid to help cool gases like oxygen and nitrogen, to such a temperature that they would liquefy. Commercial use to keep liquids hot was a successful by-product of this superb product.

Waterbed 1832

Physician Neil Arnott, created a 'hydrostatic' bed. He produced this device to prevent bedsores for invalids. Its construction included a large bath full of water and a rubber coated canvas on to which normal bedding was placed. It was never patented but well documented and Arnott freely let anybody use his design in the cause of medical advancement.

The industrial revolution 1700–1900

Great Britain, underpinned by coal, steel, iron and steam became the first and leading nation in the global industrial revolution. It is often said that Britain's Industrial Revolution created the modern world. It all began in Britain in the latter part of the 1700s, when most manufacturing was a cottage industry and essential products such as nails; clothing, timber and other basic produce of the day were produced by hand or with basic machines in people's homes. Even farming was a hugely manual and laborious task and very inefficient in its output. But rapid industrialisation marked a surge towards purpose made powered machinery and specialist factories designed solely for mass production. The key driver behind this rapid development was the textile and iron industries that played a pivotal role in the rapid growth, innovation and impetus in the Industrial Revolution.

Supporting industries played a key role in the success of the Industrial Revolution, seeing huge developments in banking, Communication and transport systems. It was clear that such a huge cultural step change could not be achieved if the nation's people did not benefit. For the first time in Great Britain's history, this rapid industrialisation not only generated considerable variety of manufactured goods to the people, but it also provided improved living standards for many, taking people from abject poverty and improving living conditions for the working classes and the poor.

Prior to the Industrial Revolution, the majority of people lived in very small rural communities, centred largely on farming. Life was not easy as low incomes meant that many people had to make their own clothes, furniture, tools and grow their own food. But this was all set to change. Industrialisation attracted workers from rural communities with regular, higher wages, farming became mechanised, so fewer workers were required on the land and small towns centred around factories and mills were starting to grow at unprecedented speed into huge industrialised cities.

A number of factors contributed to Great Britain's role as the founder and birthplace of the industrial revolution. Firstly, it had huge deposits of iron ore and coal, essential for industrialisation and British society was politically stable. Secondly, Great Britain was the most powerful colonial power at the time. This meant that it had an instant market for its manufactured goods and its colonies could also provide much-needed raw materials to power the enormous growth and demand in industrialisation.

The rise of modernisation and mechanisation was fuelled by the demand for British goods. A generation of innovation and invention developed through necessity as industrialisation spread across the land. Eventually, other countries witnessed the growth in wealth and saw how industrialisation could benefit them too. So the British introduced laws prohibiting the export of skilled workers and their valuable technology. However, these laws could not prevent the spread of industrialisation to other countries, particularly, Germany, France, Holland and the United States of America.

By the mid 1800s, there was a well-established industrial base throughout the North-Eastern seaboard of the USA and most of Western Europe. Such was the speed of industrialisation in other countries that Great Britain lost its place in the rankings of industrialised nations by the 1920's and the USA took its crown.

Manufacturing revolution

Through the course of history, the process of manufacturing has developed spectacularly. Producers of goods have created methods to mechanise their systems and as a consequence reduce labour costs and speed production in the process. This transformation in production began in 18th century Great Britain and eventually spread to mainland Europe.

By the late 18th Century it had spread to the shores of America. The main benefactors of these huge changes were the industries of agriculture, mining, goods manufacturing, glass making, military hardware and textiles. At the height of the industrial revolution, Britain was the largest manufacturer by volume, in the world. Despite the small size of the country, Britain is still the world's 6th largest manufacturer.

Sewage systems 1865

As the chief engineer in the mid 1800's for the Metropolitan Board of Works (LMBW) in London, Joseph Bazalgette made a huge impact on the health and wellbeing of the population of the capital through his innovative and efficient development of its large bore underground sewage system. Headed by Bazalgette, the establishment of the LMBW meant that for the first time in the city's history, a single organisation was able to plan and co-ordinate public civil works in a holistic way across the whole city. An outbreak of cholera prompted the city heads to power forward with a plan to begin works on street drainage and sewer improvements.

By the end of 1866 most of London was connected to a state of the art deep sewer system that is still being used to this day. Bazalgette ensured that the new system of low level sewers diverted all foul water and connected new and existing sewers to a network of riverfront treatment works. Thus, ensuring for the first time in the city's history, that no raw or untreated sewage was pumped into the River Thames. His legacy and inspiration lived on long after his death in 1891, with places as far afield as Mauritius, Port Louis and Budapest enjoying the benefits of his advice, design and planning.

> 'In the industrial revolution Britain led the world in advances that enabled mass production: trade exchanges, transportation, factory technology and new skills needed for the new industrialised world'
>
> Lucy Powell

Blast Furnace

Industry – process and innovation

Aluminium the discovery of 1808

Humphrey Davy identified the existence of the base metal alum, which he first named aluminum (an old English term still used in the USA) but later he renamed it aluminium. 17 years later the metal discovered by Davy was first produced in a very basic and impure form by Danish chemist Hans Christian Orsted.

Batteries Structural 2006

British company BAE systems developed an innovative method of building batteries into vehicle or body armour that also acts as an integral element of the structure. Designed initially to lighten a soldier's load on the battlefield, this technology is now used in many electric vehicle designs

Bessemer converter Mass production of steel 1850

In the 17th and early 18th century, steel was very expensive to produce, using very inefficient power-hungry methods. However, in 1850, British inventor Sir Henry Bessemer was working on finding an inexpensive process for the mass production of steel. The key part of the process is to remove all impurities from iron by air oxidation blown through the iron in its molten state. This process in turn raised the temperature of the mass of iron, maintaining its molten state – now known as 'the Bessemer process'.

American, William Kelly around 1851, also independently improved on this process but after filing for bankruptcy, he was forced to sell his patents to Henry Bessemer. This additional information helped Bessemer fast track and refine a similar process that Kelly had also been working on, culminating in Bessemer's patent granted in 1855, for what was now a fully tried and tested process that was to change the world in steel production and use for the masses.

Blast furnace World's first coke fired 1709

Abraham Darby was an iron ore smelter who pioneered the use of coke to use in his blast furnaces. Because coke had fewer impurities, the iron Darby produced was of superior quality to iron used in a standard coal fired blast furnace.

Carbon fibre 1963

This fantastic super strong, lightweight material is only one of many inventions that were developed by British military researchers. Today Carbon Fibre has numerous applications ranging from cars, boats, sports equipment, record player pick up arms, bikes and even aircraft wings and fuselages, such as the Airbus A380 and the Boeing Dreamliner.

Cement 1824

Joseph Aspdin was a cement manufacturer who was granted a patent for Portland cement in October 1824. The patent was entitled 'an improvement in the mode of producing an artificial stone'. It was within the text of the patent that he used the phrase 'Portland cement', based on its constituent ingredients, Portland stone and oolitic limestone. Prior to his invention, the most common cement used in bricklaying was a simple limestone mix. His invention was quickly adopted, first in Great Britain, and then across the world. Today, Portland cement is still the standard used for construction the world over due to its flexibility in use and temperature tolerance.

Concrete Quick-Drying. 1774

John Smeaton was a mechanical scientist from Leeds. He started life as an instrument maker but later progressed into windmill and water wheel design. But it was his work on building the Eddystone lighthouse for which he is most famous. Savage seas and fires had destroyed the mainly wooden structures of its predecessors. Smeaton had to find a more innovative solution that was durable and long lasting. After researching many materials for use as an effective mortar, resistant to the harsh elements, Smeaton eventually developed a quick-drying concrete, known then as 'hydraulic lime'. His proposal to use clay and limestone materials using his method is regarded as the world's first use of modern concrete in engineering. To complete his task of building the lighthouse on Eddystone Rocks, he had to create a crane that was capable of lifting very heavy weights safely above the height of the lighthouse.

Concrete Reinforced. 1854

Plasterer William Boutland Wilkinson, was recognised as being the first person to develop the method of building reinforced concrete. Wilkinson built a small reinforced concrete cottage and reinforced the floor and the wall with wire rope and iron bars. This was the world's first structural reinforced concrete building.

Electro plating 1840

George R. Elkington and his Brother Henry were the inventors who first patented the process of electroplating in 1840 and exploited the process commercially. The Elkington's had a virtual monopoly for many years based on their patent for an inexpensive system of electroplating. It must be noted that the Elkington's patent was based on the British inventor John Wright's design, to which he sold the rights for his process to the Elkington brothers.

Folding carton 1920

Businessman Charles Henry Foyle is recognised as being the inventor of the folding carton in 1920, making flat packed cartons that shipped with low volume, then they could be folded into strong boxes within seconds. Foyle was also the founder of the first folding box company called 'Boxfoldia'. Foyle's invention effectively created the low-cost packaging industry, as recognised today.

Float glass 1952

Alistair Pilkington is an inventor who was knighted for his ground-breaking invention of the float glass process. Acknowledged to be one of the most significant inventions of the 20th century, Pilkington's process is now the principal method used to produce glass worldwide for windows etc. Almost all glass used throughout the world today relies on the development of British Glass Engineer, Pilkington's development from 1952. It is pure coincidence that Pilkington was, at the time of his discovery, working for British company Pilkington Glass and not related to the founders.

The Float glass process was originally only able to make glass 6mm thick. It has now been refined to make thicknesses from 0.4mm up to 25mm. Pilkington discovered that when molten glass is poured constantly at 1000 degrees Centigrade into a shallow bath of molten tin, then the glass floats on the tin, spreads and forms a supremely level surface. After controlled cooling, the glass materialises as 'process polished' with virtually perfect surfaces. Over 40 manufacturers in 30 countries, producing more than a million Tonnes of glass a week have now licensed Pilkington's process.

Hydraulic press 1795

Locksmith Joseph Bramah, developed unpickable locks, but more importantly was an avid inventor. The most significant of all his inventions was the Hydraulic Press. This machine consisted of two piston cylinders, each with a different cross-sectional area and they were connected with a tube, filled with a light viscous fluid. They were connected in such a way that one piston caused the other to move at the same time. It is still one of the most significant machine design concepts used in the industrial world today. Hydraulic presses are needed for all manner of production, from tin cans to rubber processing and car bodies to plastic moulding.

Hygrometer 1820

John Frederic Daniell was a scientist who devised an instrument to measure the dew point, where dew forms on the surface of glass, then cooled by evaporation into a jar. The jar is then cooled by the evaporation of the jar of ether on its outer surface.

Lead crystal glass 1674

George Ravenscroft was a merchant and inventor from London who had no technical ability or knowledge of glassmaking. Even more remarkable is the fact that Ravenscroft invented a new purer form of crystal glass that has stood the test of time since his invention and creation. Essentially the main difference in glass production using Ravenscroft's method is that lead replaced the calcium, typical of standard potash glass. Lead glass is recognizable by its distinct ring when tapped. Ravenscroft's discovery improved the appearance of standard glass too through its purity and clarity. Lead crystal glass was the ultimate material for storing drinks, but due to the high lead content, Crystal glass has now taken its place in today's glassware sector.

Linoleum 1860

Legend has it that inventor Frederick Walton went to his shed to get a tin of paint and upon opening the lid, he noticed that a thick flexible film of congealed linseed oil had formed on top of the fluid. This provided Walton with a great idea. From that point, he knew that linseed oil could be formed into a waterproof material and that if he glued a canvas type of backing, he could market and sell it as an 'off the shelf' floor covering. Once in production and now known as linoleum, this ubiquitous material was sold the world over to millions of homes and offices.

Macadam 1820

Macadam is a method of road construction developed by engineer John Loudon McAdam in 1820. Although his creation was not new or unique, it did simplify previous systems of road building, through the use and standardisation of layers of small size aggregate or small stones, binded by a wet-mix quick drying cement material, laid on a thick bed of compressed hardcore. Not to be confused with the more advanced early 20th century invention patented by British inventor Hooley in 1901 for the 'Tarmac' road construction method now used extensively throughout the world.

Oil First oil tycoon 1851

James Young was a chemist who was fascinated when he saw oil seeping out of the shale-laden ground near his home in Bathgate. After months of experiments and research, he refined the crude oil and created what we now know as paraffin, later developing the paraffin lamp. Young also invented the world's first lubricants for industry. He had in fact not only developed the world's first shale fracking technique of oil extraction, but in the process, became the world's first oil tycoon. Young further refined his crude oil process to produce what we now know as gasoline or petrol.

Oil refinery 1851

Chemist James Young built the world's first oil refinery to support his newfound discovery of oil in the Bathgate area. Eventually producing several grades of refined crude oil, Young paved the way not only for the British oil industry, but also the world's oil industry in general. As a result, Young formed the world's first commercial oil company.

Pneumatic tyres Air inflated 1887

John Boyd Dunlop invented the first practical pneumatic tyres. Although the invention was first made in 1845 when railway engineer Robert W Thomson patented the world's first Pneumatic tyres, but at the time there was little demand for them. It was more than forty years later when Dunlop further developed pneumatic tyres to prevent his son from getting headaches when riding his tricycle. Fortunately, this time his development occurred simultaneously as the new mass-market bicycle trend was growing.

Safety fuse 1831

William Bickford was a leather merchant and inventor from Devon. He is best known for his invention of the safety fuse in the field of explosives. His first design put the main explosive into a parchment sealed tube and then joined it with a smaller parchment cartridge holding gunpowder as the fuse. Bickford later thought of a clever improvement by winding rope strands around a core of gunpowder. He then varnished the rope to ensure a steady burn up the rope from its end, eventually reaching the explosive. Bickford's improved method of safety fuse is still used to this day.

Sandpaper 1833

Sandpaper or glass paper, was first manufactured in London by British company John Oakey & Sons, a manufacturer of polishing materials and sandpaper. In Oakey's early days as an apprentice piano maker, he was taught to make sandpaper by glueing powdered glass or sand on to paper. As the head of his own business now, Oakey was able to develop better processes and adhesives. This made it possible to mass produce a cost effective and high-quality abrasive paper of differing abrasive grades and set up a new manufacturing business in London in 1833 to start full scale production. He later developed 'wet & Dry' and 'emery paper' abrasives along with a compliment of furniture polishes, silversmith's soap and plate powder. He died in 1887 and his sons John and Joseph carried on the business, eventually taking the company public in 1893. Today, Oakey Abrasives is still one of the world's most respected and leading abrasives companies.

Stainless steel 1913

It was in 1912 that researcher and steelworker Henry Brearly was given a job by a small firearms producer, to develop a substance that would extend the life of their gun barrels. During his research, he discovered a method to produce corrosion resistant steel through experimenting with the addition of varying quantities of carbon and chromium to normal steel. Legend has it that he discarded some steel that he had been experimenting with and a few months later, discovered that it still looked as shiny and bright as the day he threw the steel scraps outside. At the time, he simply named this new steel as 'rustless steel'. He first applied for a US patent in 1915 and now, stainless steel is used in everything from buildings, ship making and cars to surgical instruments.

Steam hammer 1839

Inventor James Nasmyth, created the steam hammer; a revolutionary machine that helped power the progress of the Industrial Revolution. Nasmyth had a simple concept. A hammering block was inverted and hoisted by steam to a position above the metal to be fashioned and then steam in the block was released and fell. It was so advanced that both frequency of blows and the strength of impact could be controlled accurately. Nasmyth's invention allowed for much larger forgings using heavier metals. The steam hammer was a direct predecessor of the pile driver.

Tarmac 1901

Tarmac is an abbreviation of Tarmacadam and is based on the type of road surface developed by Engineer John Loudon McAdam in 1820. However, in 1901, civil engineer Edgar Purnell Hooley patented the name and the advanced process of Tarmac using Tar, rather than cement to bind the aggregate on the road surface. Hooley then founded Tarmac PLC and it is now one of Britain's largest construction and building material groups. The advanced principle behind Hooley's development on road servicing is still the basis for all tar-based road surfacing the world over.

Water and sewerage systems 1840

William Lindley was a civil engineer born in London in 1808. Working initially with his father, then later on his own, he designed and installed the first sewage systems in Europe. Notably, Warsaw, Lodz and Prague and he designed and coordinated the water system in Baku. He died in London in May 1900.

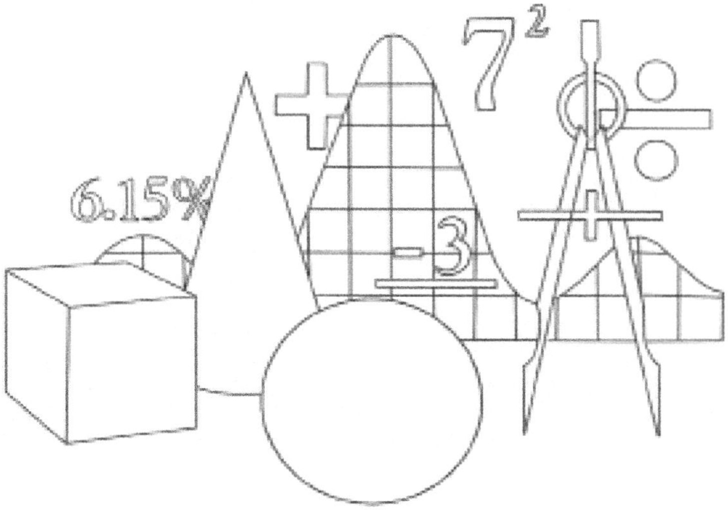

Mathematics

Boolean algebra Digital logic design 1840

George Boole, a mathematician, developed a generic technique for manipulating and representing logically valid inferences or 'mechanical ways of making transitions'. Boole's approach used unequivocally, adopting algebraic systems for this purpose.

Calculus 1669

Isaac Newton was one of two men who are independently attributed to the development of calculi and its foundations. Gottfried Wilhelm von Leibniz is the other great mathematician; although both arrived at similar conclusions, each man took a different path to reach their conceptual conclusions. Newton focused on Variables changing with time. Newton used quantities 'x' and 'y' that were limited velocities to compute the tangent. For Leibniz, the Calculus was taken towards analysis, while for Newton it was always geometrical.

Nowadays, this subject is often taught conceptually first, in a rather reverse order compared to its actual development and as a result, today scholars are often taught first about its limitations.

Equal sign 1557

Mathematician and physicist from Tenby, Robert Recorde, introduced the plus sign (+) to English speakers and created the equals sign (=), both now used universally in the world of science.

Logarithms 1614

Mathematician John Napier first discussed logarithms and published a paper of his findings in 1614 in a document entitled 'Mirifici Logarithmorum Canonis Descriptio' (Description of the Wonderful Rule of Logarithms). Napier is therefore recognised as the father and creator of logarithms.

Metric system 1668

Contrary to common belief that the French invented the metric system, it was in fact originally developed, with published findings by a British bishop called John Wilkins. Wilkins first published his system of decimal metrication, in a proposal named 'universal measure' in 1668. Wilkins developed his plan further that year and published another paper that soon became known as 'the metric system' after the Royal Society published 'An essay towards a real character and a philosophical language' and it contained quotes from Wilkins referring to his plan for a system that could be used as a 'single universal measure'. Wilkins later paper included all of the components of the international Metric unit based System. The complete naming of all the metric units came much later, but there can be no doubt that his work lay the foundations for the adoption of the metric system in the modern world.

Oil The barrel measurement 1483

The global standard for the quantity of oil and the unit by which it is measured and traded is known as the 'barrel'. The barrel measurement dates back to the reign of the English king Richard III. Between 1483 and 1484, coopers had developed a method to make barrels watertight. During this period, a Cask held 84 gallons and a Tierce held 42 gallons. The Tierce or Barrel became the standard size for carrying anything including wine, beer and fish. The 42-gallon capacity of the barrel weighed about 300 pounds and was easier to manoeuvre than the much heavier cask. British oil pioneer James Young adopted this size and design. The American Petroleum Producers association formally adopted the 42-gallon Barrel in 1872 to hold and transport oil. The 42-gallon Barrel is now firmly established as the global standard measurement for oil.

Pie chart 1854

Florence Nightingale was a gifted mathematician as well as a famous nurse, who invented the 'coxcomb chart' now known as the pie chart.

Slide rule 1622

William Oughtred was a British Anglican minister and recognised as the inventor of the slide-rule, the forerunner to the modern-day computer. In 1622 Oughtred placed two scales alongside each other and by sliding the scales back and forth to the desired point, he used them to calculate distance relationships, therefore dividing and multiplying directly without manual calculation. He is also credited with the invention of the circular slide rule. The slide rule was later used for roots, trigonometry and Logarithms by introducing a third sliding central scale.

Standard deviation Probability 1868

In the 1840s, mathematician Francis Galton, was fascinated with mechanical computing that performed both graphing and mathematical functions. Galton's greatest creation is probably his statistical correlation concepts and standard deviation. Galton was the first person to consider human differences such as intelligence, when collecting data, introducing what he called standard deviation into the output of data from questionnaires and surveys.

Venn diagram John Venn 1834 to 1923

John Venn, who was an undergraduate student in mathematics at Gonville and Caius College at Cambridge University, developed the famous Venn diagram. Venn authored many books including 'symbolic logic' and it was during this period that he developed what is now commonly known as the Venn diagram. Venn's development was driven by what he considered the inadequacies of Euler's non-intersecting diagrams or Eulerian circles because they could not show inter-relationships across three differing classes of data.

> "The study of mathematics, like the Nile, begins in minuteness but ends in magnificence"
>
> Charles Caleb Colton

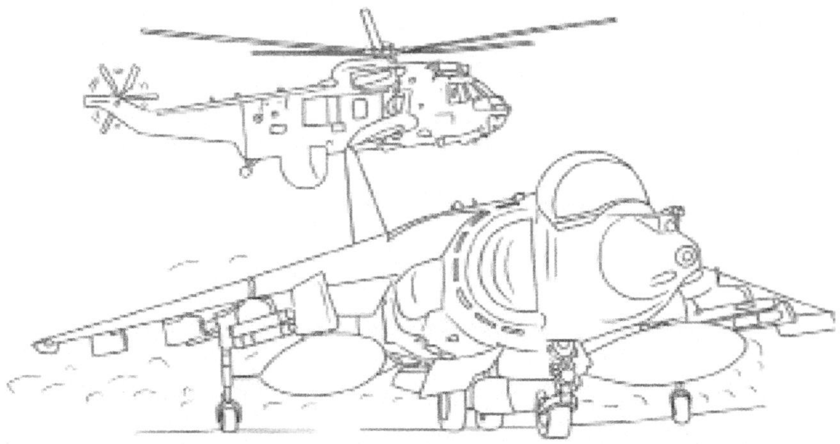

Westland Puma helicopter and Harrier VTOL

Military

Aircraft carriers (HMS Argus) 1918

In this year, naval architect James Graham's design of the world's first aircraft carrier was launched. Including all the major support systems and features such as: angled steel deck, steam catapult, tail hook, arresting cable system and aircraft command centre. It was the world's first carrier truly capable of launching and recovering naval designed aircraft.

Bouncing bomb Invention and development 1943

The bouncing bomb was designed by aeronautical engineer and inventor Dr Barnes Wallis and is now recognised as being possibly the most ingenious creation in British military engineering, even though initially, the ministries of the government regarded it as a madcap idea that would never ever work. But through his tenacity and resolve, Barnes Wallis developed his idea to reality and it was his genius that enabled the famous Dambuster raid to take place in

1943. Although the raid achieved all the goals of the plan, to destroy Germany's 'heavy water' rocket fuel plants on the Ruhr Dams by skimming a bomb above the water to avoid German Torpedo nets, 'Operation Chaste' also resulted in huge loss of British life. Resulting in 53 of the 113 aircrew lost or perished.

Depth charge 1914

Thomas Firth & Sons were a British engineering manufacturer from Sheffield. It is widely recognised that under instruction from the Royal Navy Torpedo school, they manufactured a depth charge based on Royal Naval Scientist, Herbert Taylors design, that went on to become the first useable depth charge in the world. It was designed in their words, 'for countermining as a dropping mine'. The mine was launched from a platform on the side of a warship and dropped into the water. The mines were fitted with a detonator that was actuated by water pressure automatically. If a submarine was in the vicinity, then mines of various depth setting could be launched to automatically explode at different levels. This proved hugely effective in forcing enemy submarines to surface quickly and surrender.

Gas mask 1918

Edward Harrison was a chemist from Camberwell who, through his recruitment to the World War one effort and because of his scientific background, set out to find a way of defending the troops against the onslaught of German gas weapons. Harrison's initial developments centred on his 'box respirator', the Army's first useable gas mask of which he continually made improvements to during the course of the war.

Such was the success of Harrison's mask, that his design was copied by the Germans from captured British troops and eventually formed the basis for all Gas masks the world over. Harrison was promoted from Corporal to Lieutenant Colonel for his war efforts.

Dreadnought 1906

Dreadnought was the world's first turbine powered battleship, launched in 1906 and determined the design of all future big gun warships, dominating the navies of the world for the next 40 years. This mighty ship was 526 feet long with a crew of roughly eight hundred. Breaking with tradition, this behemoth boasted 4 propeller shafts driven by steam turbines rather than traditional pistons, providing a top speed of 21 knots. Aside from its standard above deck heavy armoury, Dreadnought was built with 4 torpedo tubes to help fight off torpedo boats and destroyers. Never seeing action in battle before WW1, the Dreadnought was soon superseded by 'Super Dreadnoughts', with greater armament and better performance.

Fighter aircraft Vickers F.B.5. 1914

The Vickers F.B.5. Gunbus was a British designed and built fighter aircraft, first flown on July 17th, 1914, making it the world's first Fighter aircraft. The second crew member, or observer, sat in front of the pilot who was also responsible for aiming and firing the single 0.303 Lewis gun. It was the first aircraft specifically built for air-to-air combat.

Jet bomber 1949

The world's first jet bomber was the English Electric Canberra. This revolutionary aircraft was also the world's first high altitude bomber.

It remained in service with several air forces around the world for more than 50 years and is still being used by NASA.

Laser timing 2008

Used originally on the battlefield to prevent firing on friendly forces, this cutting-edge technology is now used in the Velodrome at Manchester to accurately time racers, designed and built by BAE (British Aerospace Engineering).

Liquid armour 2005

A team of research scientists at the British company BAE systems developed a liquid body armour that doesn't restrict movement on the battlefield nor on law enforcement personnel, by using their newly developed shear-thickening fluids (STFs). This fluid displays an instant increase in viscosity when a force is applied.

Machine gun 1718

Inventor and lawyer from London James Puckle, is credited with designing and building the world's first machine gun: 'the Puckle gun'. Puckle demonstrated his tripod mounted single barrel gun, fitted with a nine-shot cylinder that automatically revolved after each round was fired. His weapon fired at a rate of nine shots per minute, contrasting with the standard musket at the time that could be loaded and fired at best, three times a minute. Unfortunately, although well documented and demonstrated, Puckle's ground breaking weapon failed to attract investment, prompting one newspaper at the time to comment; 'those are only wounded who hold shares within'. Puckle was the first person in the world to patent what was clearly described as a 'portable machine gun".

RADAR Radio locator 1935

Sir Robert Watson-Watt designed and built the world's first RADAR capable of detecting targets and providing their true range. Watson-Watt was a British scientist who after graduating from University with a BSc, worked with Professor William Peddie. As a result of his work with Peddie, he became interested in radio waves and their application. In 1915 during WWI, Watson-Watt worked as a meteorologist for the Royal Aircraft Factory in 1915 and experimented with the use of radio waves to detect severe weather conditions to forewarn pilots of the perils that could lie ahead. In 1927 when Watson-Watt joined the National Physics Laboratory, he became an internal Superintendant of the unit.

In 1935, following many years of development and experimenting, he published a paper entitled "The detection of aircraft by radio methods". His report fascinated Sir Henry Tizard who was the head of the air defence committee. So much so, that he asked Watson-Watt to carry out what was to become a successful trial, in which short-wave radio was employed to detect an RAF bomber. Following the successful trials, Robert Watson-Watt was appointed as Superintendent of the new Air Ministry, Bawdsey research station near Felixstowe.

The ensuing research carried out by Watson-Watt and his team led to a network of RADAR stations being set up across the east and south coasts of England, proving vital in the defence of the country during the Battle of Britain. As a consequence, and for the first time ever, the RAF was given advanced warning of incoming attacks by the Luftwaffe, enabling them to prepare and react appropriately.

Watson-Watt and his team of John Randall and Henry Boot then advanced their research and developed a 'Cavity Magnetron'. This meant for the first time, a portable unit could be sited on a fighter or Bomber aircraft, giving the pilots even greater warning of the approaching enemy. Magnetrons could also heat up water and today, the same devices are to be found in every microwave in the world and used as a means of quickly heating foods or liquids. Watson-Watt was knighted in 1942 and given £50,000 for services to his country. The USA also presented him with the US Medal of Merit for his huge contribution in WWII. He died in 1973. RADAR stands for RADio Detection And Ranging.

RADAR HS2 Airborne radar 1939

Alan Blumlein was possibly one of the world's greatest electrical engineers and pioneers. This British genius invented binaural sound form a single source (Stereo as we now know it), which he patented in 1933, then in 1936 he created an HD television system that would replace Baird's crude mechanical system. But possibly his greatest contribution to the world was his invention of Air Interception (AI) Radar. Alan Blumlein was instrumental in creating and implementing the world's first operational Air Interception Radar system. This huge technological advance played a key role in helping the RAF fend off the Luftwaffe night bomber offensive.

The most important Radar project in the latter part of the war was the H2S, a technique that enabled the RAF to accurately bomb through thick cloud. However, before he was able to see the fruits of his years of development and invention come to bear, he was tragically killed when a Halifax bomber that he was travelling in to test the system, crashed in the Wye Valley, killing all 11 crew including Blumlein. His teams of engineers were able to complete this crucial project successfully and it proved decisive in shaping the outcome of the war. He died aged 38 and to this day has received no recognition or honour for his huge contribution to science and the war effort.

Rocket Solid fuel 1804

Inventor William Congreve designed and built the 'Congreve Rocket', a British military weapon. The weapon was developed for the Royal arsenal following several bloody wars fought between the Kingdom of Mysore in India and the British East India Company. The rockets were the first of their kind to be powered by solid fuel and were used successfully in the Napoleonic wars of 1812.

Special forces SAS and SBS 1941

The SAS and SBS are the world's first and leading Special Forces and were the model for the USA's special forces 'Delta Force'. SAS (Special Air Service) and SBS (Special Boat Service) were conceived and established in the early 1940s as elite regiments that could work and infiltrate behind enemy lines without detection. Admission to both elite units is by invitation only. Made up of mainly British soldiers who have to pass an exhausting selection test for acceptance and entry into the elite regiments. The SAS has become a template for most other countries' special forces. The SBS is the least famous, but equivalent of the SAS in the Royal Marines.

Ship Steam Turbine 1897

Turbinea was launched as the world's first steam turbine powered ship in 1897. Built as an experimental Royal Naval ship it was at the time, the fastest ship in the world, setting the benchmark for years to come.

Sonar 1912

Mathematician Lewis Fry-Richardson was from Newcastle-upon-Tyne and created the world's first Sonar System. Sonar is the acronym for SOund Navigation And Ranging. Sonar works similarly to Radar, but emits sound waves through the water and then bounces them back from submerged objects that lie hidden to the naked eye. Other countries have reproduced Fry-Richardson's invention. As reparation for WWII, Britain gave the technology to the Americans.

Stun grenades SAS 1960s

Britain's elite special service military unit, the SAS, developed the stun grenade. It sent a specification to the Royal Ordnance unit in Enfield North London and resulted in the 'G60 grenade' or 'flash bang' stun grenade. The stun grenade is designed to distract and disorientate the enemy in a building, providing the assault team critical seconds advantage to enter and counteract the threat. The SAS stun grenade was designed to create a powerful and loud bang exceeding 170 decibels in volume, with a blinding light flash and no lasting injury. Military forces worldwide have now implemented the stun grenade; such is its overwhelming success.

Tank Military 1914

It was whilst reporting as an official war correspondent, that Ernest Swinton devised a design for a Military tank. He suggested that if the crawler tractors that were then used to pull military armoury into battle on the Western front, were combined as a tracked cannon, then it would solve many issues, scoring a competitive advantage over the enemy. His design went into manufacture eventually in

1914 and it was the first 'Military Tank' in the world, capable of scaling a five-foot high obstacle and traversing six-foot trenches with ease. It was also armoured to protect the crew and travelled at a leisurely 4 mph.

Submarine 1578

Designed by mathematician and naval writer William Bourne, but not built until 1624 by Dutchman Cornelius Drebbel. Bourne is credited with designing and publishing a paper on the first credible submarine ship designed to be navigated underwater, distributed in 1578. Bourne put forward drawings and calculations for a totally enclosed 'underwater boat' that could be immersed and rowed underwater. It was built with a wooden frame and the body was covered in a thick skin of water resistant leather. Its design enabled it to be submerged by contracting the boats volume through hand levers inside and then reversing the function to climb back to the surface.

Torpedo 1866

Engineer, Robert Whitehead first designed and developed an underwater missile that could be launched from a ship in an underwater tube whilst at sea. His first version of 'Torpedo' was powered by compressed air rather than explosive charges and was fitted with a mechanism inside that enabled the missile to adjust and stay at a constant depth based on water pressure. It is on record that the very first ship to be sunk by Whitehead's torpedo design was a Turkish steamer called Initibah and was launched from a Russian warship in 1878. His invention notably changed warfare at sea.

"Who dares, wins.
Who sweats, wins.
Who plans, wins."

British Special Air Service (SAS)

The Beatles

Music

The Beatles 1960 to 1970

The most successful pop group in the world. The 'Fab Four' as they were affectionately known, came from Liverpool and were formed in 1960. The four original band members were George Harrison, John Lennon, Paul McCartney and Pete Best. Ringo Starr replaced Best in 1962. Their manager was Brian Epstein, who encouraged George Martin to sign them up to the EMI owned Parlophone record label.

Their sales momentum and popularity did not diminish. The Beatles sold a staggering total of more than 600 Million records during the 60's alone, clocking up scores of top 10 hits and number ones, winning 14 international awards in that decade. Statistics alone prove that the Beatles were by far the first world super band and the most successful ever.

Concertina 1829

Developed by Sir Charles Wheatstone and patented in 1829, he subsequently filed a second patent for an improved version in 1844. Probably developed separately, Carl Friedrich Uhlig announced a German version five years after Wheatstone's original patent. The concertina is a wind powered reed instrument, driven by hand held bellows and sound is generated by pushing and drawing, whilst simultaneously playing keys on either side of the hand-held bellow caps. Each end cap and bellow set is Hexagon shaped; though later variants were octagon shaped too.

Edward Elgar 1857 to 1934

Born in Lower Broadheath on the outskirts of Worcester, Elgar was one of Britain's outstanding composers with many of his compositions gaining international recognition. He is most famous for the 'Enigma Variations' and his 'Pomp and Circumstance Marches'.

English horn 1720

The English horn dates back to 1720 and is a variant of the Oboe, being roughly one and a half times longer with a pear-shaped bell and pitched in F, rather than the standard oboe's C pitch. The English horn is very popular in France and is known there as the Cor Anglais.

Glastonbury Largest music festival 1970

Created by British farmer's son Michael Eavis' with the inaugural Glastonbury music festival held in 1970, formerly known as the Pilton festival, this multi genre festival is by far the world's largest and the most famous music festival. Glastonbury has been running continually since 1970 with the exception of cyclical 'fallow years' to allow the land to recuperate. The festival's first headline act was Tyrannosaurus Rex, better known as T. Rex, and now the festival attracts music fans from all over the world, with tickets normally selling out within the first hour of release. The Glastonbury festival regularly attracts more than 175,000 festival goers with past headline acts such as Stevie Wonder, David Bowie, Rolling Stones, Dolly Parton, Muse and Coldplay.

Gustav Holst 1874 to 1934

Born in Cheltenham, Gustav Holst is without a doubt one of Britain's greatest ever composers of the classical genre. His most famous composition is the Planets Suite. He became friendly with composer Ralph Vaughan Williams in 1895, the start of a lifelong companionship. In 1905 he became musical Director at St Paul's Girls School in Hammersmith and continued to teach there until the end of his life. It took Holst two years to write his most prolific works, the Planet's, between 1914 and 1916.

Holst received perhaps the greatest accolade of his illustrious career in 1929. He was awarded the Howland Memorial Prize from Yale University for excellence in the arts and the gold medal of the Royal Philharmonic Society in 1930. He was appointed visiting lecturer at Harvard University in 1932 but became ill, forcing him to return to Britain, later dying on 25th May 1934.

Live aid 1985

Live aid was a concert born in Britain as a response from the natural disasters sweeping Africa at the time and the subsequent need to raise money to aid immediate disaster relief. It was conceived by Midge Ure, Harvey Goldsmith and Bob Geldof and after only twelve weeks of planning, evolved as the most successful one-time charity fundraiser of all time, raising in excess of one hundred and fifty million pounds. It is also recognised as being the world's largest and most successful music concert, seen by more than two billion people worldwide. In attendance across both venues were fifty-eight of the world's most prolific bands and pop artists giving their time for free. More than one hundred and seventy thousand people attended the two venues in London and Philadelphia USA. Phil Collins famously became the only act in Live Aid to play at both venues on the same day, by performing and finishing his set in Wembley Stadium, London, then jetting off to JFK, New York by Concorde to play the same song set in Philadelphia later that day.

Theatre organ 1887

The theatre organ is a unique air powered pipe organ and is often called the cinema organ. Robert Hope-Jones is considered to be the inventor of the theatre organ, conceived from a desire to simulate the entire ensemble of orchestral instruments in one device. He built 246 complete organs between 1887 and 1911 and eventually merged his business with the famous US Wurlitzer Company in 1914.

Three choir's festival 1719

The oldest music festival in the world is the 'Three choirs festival' held each August in Hereford, Worcester or Gloucester Cathedral. The festival originally hosted three choirs, one from each county Cathedral, who competed against each other over a period of a week. Now the festival attracts International soloists and major ensembles from all over the world competing for a coveted Finalists prize. Each year the festival is hosted and organised by the Cathedral Organist. This festival is closely associated with the famous British Composers Ralph Vaughan Williams and Edward Elgar.

Tuning fork 1711

Still used today, by professional piano tuners in particular, to tune genuine stringed pianos. British musician John Shore, invented a finely tuned steel fork, that when tapped with another steel object would emit a perfect pitch in a particular key through resonance for several seconds, thus allowing a piano tuner to set for instance middle C as the benchmark for all other strings. Prior to his invention, tuners relied on pitch pipes made of wood and were susceptible to pitch changes due to ambient temperature and humidity changes. The tuning fork had none of these negative characteristics, instead holding its pitch perfectly across a whole range of environmental conditions. The tuning fork is still in use today, although as in many areas of science, electronic devices have now replaced it.

Ralph Vaughan Williams 1872–1958

Born in Down Ampney, Gloucestershire, Ralph Vaughan Williams was a prolific British composer of film scores, choral music, opera, chamber music and symphonies. His commercially successful compositions number more than five hundred, including film scores such as the 49th Parallel, Scott of the Antarctic and Coastal Command and hymns such as: For all the saints and Hail thee festival day.

Yesterday Beatles. 1965

The world's most covered song is Yesterday by the Beatles according to the Guinness World Records and has been covered seven million times in the 20th century alone. Versions range from Frank Sinatra to Wet Wet Wet and in more than 100 different languages. It also holds the title for being the most played record on radio in the world.

TOP 20 British bands & singers by sales

1. The Beatles	600m
2. Elton John	300m
3. Pink Floyd	250m
4. Queen	200m
5. Rolling Stones	190m
6. Phil Collins	160m
7. Genesis	130m
8. Bee Gees	120m
9. Dire Straits	115m
10. Rod Stewart	110m
11. Fleetwood Mac	100m
12. Paul McCartney	95m

13. George Michael	93m
14. The Who	90m
15. Coldplay	80m
16. Spice Girls	75m
17. Depeche Mode	73m
18. The Police	72m
19. Duran Duran	71m
20. Oasis	70m

'If music be the food of love… play on'.

William Shakespeare

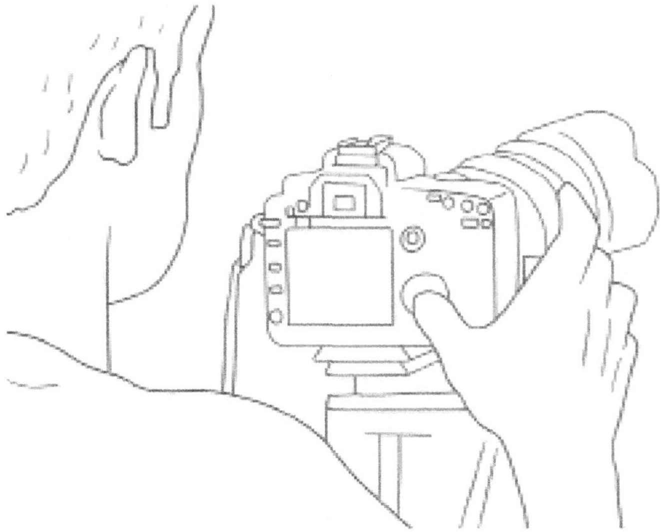

Photography and Cinematics

Cartoon First moving cartoon film 1900

J Stuart Blackton was a film producer. After travelling to the USA, he is recorded as being the first person to sketch and direct the world's first cartoon film called 'The enchanted drawing'. This was a ground breaking silent cartoon movie, famous for containing the first animated sequences recorded on standard picture film.

Cinematography Creator 1889

William Friese Greene was a portrait photographer and inventor. He built and operated studios in both Bristol and Bath and later added two more, one in Brighton and the other in London. He met John Rudge who had developed a 'Magic Lantern' that was unique in the way it could display seven slides in swift progression, giving the effect of movement. Friese-Greene was enthralled by what he had seen and was soon working with Rudge to develop his Lantern so that it could project photographic plate images on to a wall. They named this device a 'Biophatascope' and realising that plates had huge limitations, started experimenting with oiled paper and then progressing to celluloid film.

On June 21st 1889, Friese-Greene was issued with a patent for his new 'Chronophotographic' Camera. Capable of shooting 10 frames per second using perforated celluloid to enable a cog either side of the film, to turn the reel of film by a handle. However, due to the low frame speed and the unreliability of the device it did not gain the popularity he had hoped. He later experimented with stereoscopic moving images and later in the 1890s, colour moving images. His 'hobby' of cinemagraphics had unfortunately drained funds from his other businesses and he was officially declared bankrupt in 1891. Friese-Greene was widely regarded as being 'the father of cinematography'.

Kinemacolour Colour movie process 1906

Inventor George Albert Smith invented Kinemacolour. It was the world's first commercially successful colour motion picture system and started to be used commercially in 1908. The Kinemacolour process was effectively unchallenged until 1914. Smith came from Brighton and was hugely influenced by the day's great pioneers such as Edward R Turner and William N L Davidson during the process's development. Charles Urban of Urban's Trading Company in London, marketed Kinemacolour on Smith's behalf.

From 1909 it was rebranded as Kinemacolor, because Urban thought the name would be better accepted in lucrative American market. Kinemacolor used a two-colour additive process, by filming and projecting a black and white film behind alternating green and red filters. The film had to be shot at 32 frames per minute rather than the standard 16fpm and was fitted with a spinning colour filter as well as a standard shutter. The filter consisted of four equal sections, two filled with green dyed gelatine two filled with red dyed gelatine, so that the four colour combination was red, green, red, green. Panchromatic film was used and the negative processed normally. There was actually no colour in the film itself.

The first successful screening using Kinemacolour was called 'A visit to the seaside'. It was shown in Brighton in September 1908. Such was its success, that more than 300 cinemas throughout Great Britain installed Kinemacolour projectors. More than three hundred major films were shot and filmed using the Kinemacolour process.

Photography Practical 1835

It is very difficult to say exactly who invented photography. Certainly in 1826, we know that Frenchman Joseph Niepce created the first fixed image that took more than 8 hours to expose. However, what made British inventor William Henry Fox Talbot's invention ground breaking in the world of photography, was that by coating paper with silver oxide he had discovered a method of producing a translucent negative that could be used again and again to make any amount of 'positive' photographs through contact printing. This became a common method used from this point on until the more recent arrival of digital photography.

Stereoscope 3D 1838

Sir Charles Wheatstone was a physicist, born in Gloucester in 1802. Wheatstone developed a device for observing pictures in three dimensions. He called the instrument a mirror stereoscopic viewer that required two pictures of the same subject to be reversed laterally. The benefit of this arrangement was that it was not just limited to small images. This is why his stereoscope principle is still in use today for screening X-Ray 3D pictures and aerial images.

Publishing and authors

Agatha Christie 1890-1976

Agatha Christie was a crime novelist and playwright. Guinness Book of Records notes that she was and still is the best-selling novelist of all time, with her novels selling more than 4 billion copies worldwide. She also wrote under the pseudonym of Mary Westmacott and was credited with six romantic novels. Born into a wealthy middle-class family in Torquay, she is best known for her 66 crime novels and 14 short stories written under her own name.

Most of Christie's crime novels revolve around the characters of Hercule Poirot and Miss Jane Marple. Finally, she is also credited with writing the world's longest running play; 'the Mousetrap' that first opened its curtains in 1952 and is still running to date. Its 50th anniversary was held on 25th November 2002, whilst its 25,000th performance was held on 18th November 2012.

Anne Brontë 1820-1849

Novelist, Anne Brontë was the youngest daughter of the literary Brontë family. She spent most of her life in the Yorkshire moors, living with her family. After leaving boarding school at the age of nineteen she went on to become a School governess and then a teacher. After her teaching career, she became a writer of poems and novels. Her most notable novels being: Agnes Grey and The Tenant of Wildfell Hall.

Arthur Conan Doyle 1859-1930

Sir Arthur Conan Doyle was a novelist who is mainly famous for his many books on the fictional private detective, Sherlock Holmes.

Beatrix Potter 1866-1943

Born in Kensington, London, Potter was schooled by a private governess and had infrequent contact with other people. She loved drawing and studying her pets, becoming the inspiration for her many children's novels. Her parents often visited the Lake District for the three summer months of the year and rented Wray Castle, Ambleside. Over her many visits to Ambleside, Potter was inspired by her love of the countryside and the animals within, leading eventually to her writing and publishing her first book: The tale of Peter Rabbit in 1902. Other novels of note were: The Tale of Benjamin Bunny, The Tale of Squirrel Nutkin and The Tale of Jemima Puddle Duck.

Cambridge University Press 1534

Operating continuously since 1584, Cambridge University Press (CUP) is the oldest publisher and printer in the world. CUP is the second largest University press in the world next to Oxford University Press, which is the world's largest.

Charles Dickens 1814–1870

Dickens was a prolific author of some of the most creative and enduring literary classics. Dickens books are read throughout the world as staple English Literature, such as: The Old Curiosity Shop, A Christmas Carol, Oliver Twist, A Tale of Two Cities, Great Expectations & The Pickwick Papers.

Charlotte Brontë 1816-1855

Poet and novelist, Charlotte Brontë, one of the famous three Brontë sisters and is noted for her brilliant novel's such as: Jayne Eyre, Shirley, The Professor and Villette.

Cheltenham Literature Festival 1949

Founded in 1949, the world's longest running literature festival is The Cheltenham Literature Festival, now sponsored by the Times and Sunday Times newspaper group. The festival attracts some of the most prominent literary, film & TV stars from around the world including: JK Rowling, Salman Rushdie, Ruth Rendell, Stephen Fry, Dame Judi Dench, Dr Stephen Hawking, Bob Geldof, Rupert Everett, Roger Moore, Ian Rankin and Hilary Clinton.

Daniel Defoe 1660-1731

A writer, journalist, spy and trader, Daniel Foe was born in London. More commonly known by his pen name; Daniel Defoe, he is renowned for his novel's Robinson Crusoe, Moll Flanders, The Farther Adventures of Robinson Crusoe and Colonel Jack.

D.H. Lawrence 1815-1930

David Herbert Lawrence was an author widely acknowledged in shaping 20th century fiction writing. His most famous and controversial novel was Lady Chatterley's Lover in 1928.

Dictionary 1755

Dr Samuel Johnson was a scholar who published the world's first English language dictionary, simply entitled 'A Dictionary of the English Language'. It is for this reason that it is regarded as one of the most influential dictionaries in the history of the English language. There were limited forms of dictionary produced before Johnson's publication, but Johnson's was the first to focus on words used in everyday conversation aimed specifically at the layman. It was not until 173 years later when the Oxford English Dictionary was published, that a competitor was of the same standard and detail as Johnson's original work.

Emily Jane Brontë 1818-1848

A poet and novelist and one of three famous literary sisters, Emily Brontë was born in Yorkshire in 1818. She is remembered mostly for her one and only novel: Wuthering Heights. This masterpiece is now recognized as a Classic piece of English Literature.

Enid Blyton 1897-1968

Born in London, Enid Mary Blyton was a prolific children's writer whose books have sold more than 600 million copies. They include such classics as: The Famous Five, Noddy, The Secret Seven and The Enchanted Wood.

English book First printed 1475

The Recuyell of the Histories of Troye was the very first book to be printed in the English language by William Caxton, printed and published in 1475. It was translated by Caxton from French to English and took him almost 4 years to complete, due in part to his poor grasp of the French language, but was later helped and encouraged by Margaret Duchess of Burgundy. After setting up his printing press in his Westminster home in London, he proceeded with the onerous task of translating and printing several other notable French works, all published in English.

George Byron 1788-1824

George Gordon Byron was a poet, who inherited his great Uncles title of 'Lord Byron' at the age of ten. Byron was famous for such books as: Don Juan, Manfred and The Giaour.

George Elliot 1819-1880

Mary Anne Evans was best known by her pen name as George Elliot. She was a Victorian translator, journalist and novelist. George Elliot was best known for the novels: The Mill on the Floss, Middlemarch, Silas Marner and Daniel Deronda.

George Orwell (Eric Arthur Blair) 1903-1950

George Orwell was a novelist born in Bengal, India to parents who attended the world-famous Eton College. His most famous novels include, Animal Farm and Nineteen Eighty-Four.

H.G. Wells 1866–1946

Herbert George Wells was an English author, most famous for his science fiction writing. Notable works include; The Time Machine, The Invisible Man, War of the Worlds, The First Men and When the Sleeper Wakes. In the literary world, Wells was widely regarded as 'The father of science fiction'.

Kingsley Amis 1922-1995

Born in London, Kingsley Amis was a poet, teacher and novelist, best remembered for the novels: Only Two Can play, Lucky Jim, The Green Man and Take a Girl Like You.

Ian Fleming 1908–1964

Ian Fleming was an author, journalist and naval intelligence officer. Most famous for his novels based on the fictional British spy; James Bond. Having sold more than 100 million James Bond books; the vast majority turned into worldwide film blockbusters. He also wrote the famous novel: Chitty-Chitty-Bang-Bang, later made into a major global film blockbuster and play.

Jane Austin 1775-1817

Born in Steventon, Hampshire, Jane Austin was a novelist famous for romantic fiction based around the English upper class and is famous for such literary classics as: Pride and Prejudice, Becoming Jane, Sense and Sensibility and Mansfield Park.

J.K Rowling 1965-

World famous author of the Harry Potter novels was born in Yate, near Bristol, J.K. Rowling has sold more than 400 Million books worldwide. The film adaptation of Harry Potter became the world's most lucrative and successful film franchise in cinematic history.

Rudyard Kipling 1865–1936

Rudyard Kipling was an author born in Bombay (now Mumbai), India, to British parents and educated in Britain. Most famous for his novel The Jungle Book, more recently (20th Century) made into a blockbuster classic cartoon by the Disney Corporation.

J.M. Barry 1860–1937

Author and dramatist J.M. Barry, is remembered most for his novel Peter Pan, later turned into a blockbuster movie by Disney. The name Wendy was first mentioned in this story and has gone on to be a hugely popular girls name ever since.

Lancet First medical journal 1823

With the first edition of 'The Lancet' published on 5th October 1823, the publication became the world's longest running medical journal, still produced in its thousands today. Thomas Wakley was a British Member of Parliament and medical coroner who started the Lancet publication in partnership with James Wardrop, William Lawrence and William Cobbett. By 1830 it boasted a circulation of more than 4000 copies. The Lancet was, at times, very controversial, often exposing bad practice in the medical profession, to such an extent that over its many years in publication, it is probably one of the single biggest drivers of improvements in the medical profession and its practice.

Newspaper First printed in Braille 1892

The first newspaper to be printed in Braille was the British publication 'The Weekly Summary'.

Roald Dahl 1916-1990

Born in Britain to Norwegian parents, Roald Dahl rose to the senior rank of Wing Commander in the Royal Air Force as a successful fighter pilot before embarking on his hugely successful career as one of the world's best-selling authors, in both the adult and children's genre. He was responsible for some of the biggest selling books in the twentieth century literary world, including: Charlie and the Chocolate Factory, Matilda, James and the Giant Peach, BFG, Tales of the Unexpected and Fantastic Mr Fox. All were turned into best-selling movies or TV dramas.

Robert Louis Stevenson 1850-1894

Born in in Edinburgh, Robert Louis Stevenson was a poet and novelist most famous for his works: Treasure Island, Kidnapped, Strange Case of Dr Jekyll and Mr Hyde and The Master of Ballantrae.

Shakespeare 1564-1616

William Shakespeare is recognised as being the world's greatest and most successful playwright and dramatist. Shakespeare was born in Stratford-upon-Avon in the county of Warwickshire. His Father – John Shakespeare, was a local glove maker and wool merchant. His mother Mary Arden, was the daughter of an upper-class landowner from Wilmcote.

Shakespeare was educated at nearby King Edward VI Grammar School in Stratford. He married young at 18 to the daughter of a local farmer, Anne Hathaway. She was 26 when they married and Susanna was born 6 months later after their wedding. They had twins, Judith & Hamnet two years later, but their son died aged 11 years.

Shakespeare started his professional theatre career as an actor and playwright and was one of the managing partners of the Lord Chamberlain's company, subsequently known as the Kings Company after James succeeded to the throne. The company bought two theatres in the Southbank area of London; the Globe and Blackfriars. Shakespeare's first published poems 'The rape of Lucrece' and 'Venus and Adonis' were published between 1593 and 1594. As a playwright, Shakespeare was prolific. His first plays began to materialise in 1594 and he continued to average two each year up to 1611.

He began to prosper soon after his first publication with his plays, mainly histories and comedies such as 'A Midsummer Night's Dream', 'Richard II', 'Romeo and Juliet', The Merchant of Venice' and 'Titus Andronicus'. Towards the end of Elizabeth I's reign, Shakespeare was well established as a famous poet and playwright. Moving from London, he spent the final years of his life in Stratford-upon-Avon.

Shakespeare passed away on 23rd April 1616 at the age of 52 and was buried at Holy Trinity Church in Stratford, having left a legacy of 37 plays in all, that are now taught in schools internationally as a staple in the subject of English Literature.

Shorthand 1837

Author, Sir Isaac Pitman developed a shorthand method of writing, in the phonetic system to become the main shorthand method in the English-speaking world. Symbols signify sounds not letters and words and by and large are written in shorthand as they are spoken. Other shorthand systems followed, but Pitman was the most popular in the UK and second in the USA.

Stereotyping 1725

William Ged was a goldsmith who invented a method in which a whole page of printing type could be cast as a single mould, better known as Stereotyping.

Timetable National 1839

British engraver and printer George Bradshaw compiled and published 'Bradshaw's Railway Time-tables' and it became the world's first national railway timetable. Bradshaw took great care with the accuracy of his small, cloth bound book and travelled most of the routes himself, later publishing a book of his findings to offer his opinion of each location and station on the British rail network. This was probably the first national travel guide, certainly in the UK.

Thesaurus 1852

Peter Mark Roget was a London born lexicographer, natural theologian and physician who published his 'Thesaurus of English Words and phrases', better known as Roget's Thesaurus. Upon his retirement from medicine, Dr Roget started his life ambition of compiling a reference book that would contain the most used English Phrases and English Synonyms, with the aim of stimulating ideas and assisting in literary arrangement and composition. His book Roget's Thesaurus is still the most famous and continuously published book of Synonyms in the English-speaking world.

Thomas Hardy 1832-1898

Born in Dorset, Thomas Hardy was a Victorian poet and romantic novelist. His most notable works include: Far From the Madding Crowd, The Mayor of Casterbridge, Jude the Obscure and The Return of the Native.

Typewriter 1714

Henry Mill was granted the world's first patent for a typewriter. Mill was an inventor who worked as an engineer at a local waterworks. The device he invented looked to be similar to a standard typewriter of the 20th century, although he described it as a 'machine for transcribing letter'. Nothing more is known, but it is certain that further development by engineers and inventors through the centuries used his designs as a basis for the modern typewriters of the 20th century.

Virginia Woolf 1882-1941

Born in London, Virginia Woolf was a novelist of the modernist genre. Her most famous works include: Mrs Dalloway, The Waves, A Room of One's Own, and To the Lighthouse.

Wendy 1904

The first girl in the world to be named Wendy was Wendy Darling in the J.M. Barry play Peter Pan, later turned into a book in 1911. Since the play and book were published, Peter Pan has become a staple character in children's libraries.

> "I don't think necessity is the mother of invention. Invention arises directly from idleness, possibly also from laziness. To save oneself trouble."
>
> Agatha Christie

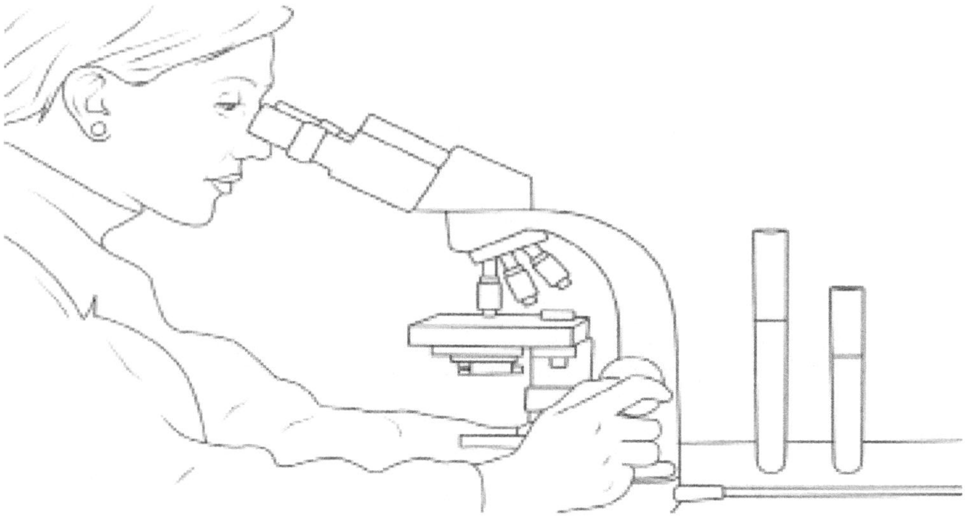

Science

Atom Discovery of 1911

Ernest Rutherford was a New Zealand born British physicist, who through his ground-breaking research work attracted the unofficial title of the father of nuclear physics. Rutherford discovered the concept of radioactive half-life, by proving that radioactivity involved the metamorphosis of one chemical element to another. He also noted differentiation of atoms and named Alpha and Beta radiation. In 1907, he moved what is now known as the University of Manchester, where he worked with Thomas Royds, proving that alpha radiation is in fact helium ions. In 1911, Rutherford made his most significant discovery where he theorised that atoms concentrate their charge in a very small core or nucleus, although he couldn't prove whether it was positive or negative at the time. This led to him establishing the Rutherford model of the atom, through 'Rutherford scattering' in his famous gold foil experiment. He is also widely recognised with the first 'splitting of the atom' in 1917 through a nuclear reaction between nitrogen and alpha particles, in which he discovered and named the proton.

Atomic theory 1805

Chemist John Dalton came a from a Quaker background, which may have contributed to his modesty that ultimately prevented him accepting fame and recognition for his pioneering work that we know today. He is known mainly for his work on atomic theory and although this work is now more than two centuries old, Dalton's theory still holds valid in modern chemistry and teachings to this day. Dalton's theory states that:

1) All matter is made of atoms. Atoms are indivisible and indestructible.

2) All atoms of a given element are identical in mass and properties.

3) Compounds are formed by a combination of two or more different kinds of atoms.

4) Chemical reaction is a rearrangement of atoms.

Modern atomic theory is now of course a little more complex than Dalton's original theory, but the fundamental nature of it remains valid. The main advances in knowledge mean that we now know that atoms can be destroyed through nuclear reactions but not via chemical reactions. There are also differing atoms (by their masses) within each element and these are now known as isotopes. His theory though, rapidly became established as a theoretical basis in chemistry worldwide.

Cathode ray tube 1879

Sir William Crookes was a renowned scientist, who led the way for many great discoveries. His development and research on a range of vacuum tubes was pivotal in his development of the cathode ray tube. On 22nd August 1879, he demonstrated several different vacuum tubes and discussed plasma, as 'the fourth state of matter'. Many of Crookes' discoveries based around his tubes laid the foundation for further discoveries such as the 'Braun tube' and the 'X-Ray' tube. The Braun tube developed later into the Television Tube.

Clouds Categorisation of 1802

British manufacturing chemist and amateur meteorologist from London, Luke Howard, was widely accepted to have created the categorisation and put names to cloud formations, in his proposal and presentation to the Askesian society in 1802, now standardised and accepted worldwide. He observed through his research, that clouds were grouped into distinct types and that each cloud category had a different cause and effect on the atmospheric weather conditions. He noted that clouds were evident pointers into the operation of these causes. He was the first person to observe and classify each cloud type in universal Latin.

Geological timescale 1931

Geologist Arthur Holmes, first proposed a geological timescale, soon after the finding of radioactivity and its use. In his proposal, Holmes calculated that the Earth was roughly four billion years old, much older than had been previously thought.

Maritime clock Marine chronometer 1720's

For hundreds of years of maritime adventure and discovery, one problem always hampered accurate navigation at sea and that was known as the 'longitude problem'. For every 15 degrees that a person travels east, then the local time progresses one hour ahead and when travelling west, the problem is reversed.

John Harrison was a joiner from Lincolnshire with little formal education, who sought to find a solution to this age-old problem that would transform and prolong the opportunity of safe and secure long-distance sea travel. Such was the problem to mariners that the British Government offered an award of £20,000, around £3million in today's money, for the first person to successfully solve the problem. Harrison made a total of 5 versions of his 'marine chronometer' during several years of design and research.

Harrison's main issues were that of maintaining clock accuracy in a moving ship and how to counter the differences in humidity and temperature in different climates. After many years of frustration, trials and demonstrations to various government committees to prove his worth, he eventually had the backing of parliament and academics alike, having countered many setbacks along the way. He received the first stage payment of £10,000 but had to personally plea with King George III to intervene with the government to get his final award payment. One of the first people to use his maritime clock was Captain Cook on his voyage of discovery. Cook described Harrison's invention as 'our faithful guide through all the vicissitudes of climates'.

Plastic 1855

Alexander Parkes was a metallurgist and serial inventor from Birmingham, who without help invented fully synthetic plastic. Parkes called his new material 'Parksine' and introduced it as one of the most significant building materials ever invented. He did not get rich from his creation, but out of all the inventions Parkes devised, plastic is now one of the most common building and fashioning materials ever invented.

Plasticine 1899

William Harbutt was a teacher of art, born in North Shields and worked as the headmaster for the Bath School of Arts. In the course of his work at the school, Harbutt developed a new type of modelling clay due to the limitations of natural clay. Critically, Harbutt's new modelling clay was non-toxic and did not dry out like normal clay and in use it was very pliable. He immediately saw the potential of his invention and applied for a patent that was granted to him in 1899 He called this new clay, Plasticine and it went on sale in 1900 as a malleable, more colourful alternative to modelling clay.

Polyethylene 1898

Eric Fawcett & Reginald Gibson were two British industrial chemists working at the ICI (Imperial Chemical Industries) plant in Northwich. Although discovered by accident in 1898 during an experiment, it was never developed or commercialised. In fact, it was not until 1933 that industrial practical polyethylene synthesis was discovered, once again by accident. Polyethylene formed overnight from a leaking test vessel and Fawcett and Gibson were convinced there was an abundance of uses for this substance, prompting them to obtain patents and eventual production. Due to its many top-secret uses during the war years, it was not until the 1950s that polyethylene became common.

Satellites Geostationary 1945

Sir Arthur C Clarke was a science writer, science fiction author, television host, inventor and undersea explorer. Clarke also had a talent for incredibly accurate predictions of future technology that went from the realms of science fiction to become reality with uncanny accuracy. One such prediction was in his famous proposal of geostationary Satellite communications that was published in 1945 in the Wireless World magazine. Nobody took this seriously at the time, but it actually became reality within 20 years, culminating in the launch of Intelsat 1 Early Bird on April 6th 1965, becoming the world's first commercial geostationary communication satellite.

Seismograph 1896

It was a devastating earthquake that shook Japan, leading geologist John Milne to invent and develop the seismograph. He relocated to Japan in 1875 as professor of geology and mining, and it was the devastation of the much-documented earthquake in the city of Yokahoma during 1880 that spurred Milne to start many years of intensive study into the causes and effect of this great natural disaster. It was during the post-quake period when Milne, born in Liverpool, constructed the world's first purpose built laboratory to study earthquakes.

After years of thorough research and dedication into pre-quake tremors and aftershock phenomena, Milne unveiled his quake detection machine, called the seismograph, developed with the help of his fellow British scientists, Thomas Grey and James Alfred Ewing. The seismograph detected and measured vibrations in the ground and transcribed them on to photographic paper.

Silicone 1899

Frederick Kipping was a chemical scientist from Manchester who received a PhD from Munich University after studying closed carbon chains. Kipping devoted the majority of his time studying organic silicon compounds and published fifty-four papers on the subject in the period 1899 to 1937. However, he failed to understand the huge commercial potential of his research. The opportunities that Kipping's research had provided were realised by Corning Glass, who with Dow Chemicals formed the Dow-Corning Corporation and in 1943 commenced the manufacture of silicone polymers, now used the world over, primarily as sealants.

Splitting the atom 1932

Physicist John D Cockcroft and Irish physicist Ernest Walton were pioneers in their discovery of how to split the atomic nucleus through the use of a self-built high Voltage device to accelerate the particles to high energy and high speeds. Cockcroft and Walton were influential in nuclear power development.

Spectacles Modern side arm type 1730

Innovator Edward Scarlett invented the world's first modern glasses featuring rigid side arms. Up to this point, early spectacles were held on to the wearer's face by hand or tied with silk or cotton ribbons. Scarlett's breakthrough came in 1730 when he developed rigid side arms that held the spectacles on a wearer's face, resting on the top of the user's ears.

Sunglasses 1752

Inventor James Ayscough started to experiment with tinted lenses in spectacles in the mid 1700s. He was convinced that tinted glasses with a green or blue tint could correct certain sight problems. He was not concerned at the time about protecting people's eyes from the sun; instead it wasn't until 1752 that he found this to be a useful by-product of the tinted lenses, after realising that his original theory had no vision correcting value whatsoever.

Synthetic rubber Isoprene 1891

Sir William Augustus Tilden was an inventor who produced a new synthetic rubber material from his experiments using the method of destructive distillation of turpentine. Tilden also gave isoprene its structural formula CH2=C(CH3)–CH=CH2.

Water desalination 1627

Francis Bacon was a prolific inventor among other talents. He experimented with seawater desalination, by trying to remove salt particles using a basic sand filter. It was not very effective, but it did lay the foundations for deeper investigation into this ground-breaking research.

> "But anyone who is practically acquainted with scientific work is aware that those who refuse to go beyond fact, rarely get as far as fact; and anyone who has studied the history of science knows that almost every great step therein has been made by the 'anticipation of Nature,' that is, by the invention of hypotheses, which, though verifiable, often had very little foundation to start with."
>
> Thomas Henry Huxley

Astronomy

Binary stars 1664

Physicist and astronomer Robert Hooke discovered what is known as the binary star system, Gamma Arietis in the constellation of Aries. The significance of this star system is that it is of a double star type and the orbital period of the two stars is more than 5000 years.

Black holes 1974

Physicist Stephen Hawking stimulated discussions and real recognition in a paper, suggesting that black holes existed in space, even though they appear to break two basic rules of physics; quantum mechanics and Einstein's law of relativity. His paper discussed the 'event Horizon', the area that's considered to exist around a black hole from which nothing can escape. He also proposed that black holes were losing mass and that they would eventually evaporate.

Halley's comet 1705

Edmond Halley was an astronomer who inspected and researched reports of a comet nearing Earth in 1531, 1607 and 1682. Halley concluded that the three comets were in fact the same comet nearing earth on its return orbit. He predicted that the same comet would appear again in 1758. Halley's calculations proved that some comets do orbit the sun. Although Halley did not live to see the comet's return, his discovery led to the comet being named after him.

Neptune Discovery of 1847

John Couch Adams was a scientist from Launceston who, through his passion for mathematics, proved that there must be another planet circling the sun like Earth. He published his conclusion and named the planet Neptune. Adams predicted the existence of Neptune due to irregularities of the orbit of Uranus and decided to investigate if there was another planet behind it.

Pulsars Discovery 1967

During the course of her PhD at Cambridge University, scientist Jocelyn Bell discovered an extraordinary signal at a wavelength of 3.7 metres that she felt was unusual because it was transmitting sharp bursts of radio energy regularly at about one burst a second. This phenomenon was unlike any other signals from known cosmic sources such as solar wind, galaxies or stars. Bell checked her findings with her advisor, radio astrologist Anthony Hewish. They both confirmed that these remarkable findings were synchronized with star time rather than Earth time, suggesting that the signal was definitely extra-terrestrial, not Earth sourced.

Hewish and Bell maintained the tracking of their discovery and eventually, after finding a second unrelated signal pulsing at 1.2 seconds, announced their findings but they were still not certain of the source. This prompted further research in other related areas by respected academics. Ultimately British academic Bell was awarded her PhD and British lecturer Hewish was awarded the Nobel Prize for the discovery of pulsars.

Uranus Discovery 1781

Uranus has of course been seen many times and incorrectly described as a star prior to being correctly identified as a planet of our solar system. However, it took astronomer William Herschel, who was in the back garden of his home in Bath, to incorrectly report it as a comet. Then after being involved in a chain of sightings using a telescope of his own design, he not only compared it to a comet because it had changed its place, but also noted its characteristics and similarity to that of a planet. Later that century, Hershel's observations were confirmed by other astronomers and by 1783, it was established that it was a primary planet of our solar system. Hershel initially named the planet after the King: George's planet. But the name was not popular outside of Britain. Erik Prosperin, a Swedish astronomer, suggested the name Neptune, but the influential astronomer of his time, Johann Elert Bodes choice of name – Uranus, named after his close friend's discovery of Uranium was widely adopted and accepted by 1850.

Evolution

Biology

Dinosaur Nomenclature 1856

Richard Owen was a palaeontologist and biologist who is recognised as being the first person to use the phrase Dinosaur from the Latin word 'Dinosauria' meaning 'terrible reptile', in a paper that he wrote describing fossils.

Evolution By natural selection 1859

Charles Darwin was a scientist from Shrewsbury, who created the basis of the theory of evolution through natural selection, transforming existing concepts of the natural world. Prior to Darwin's publication, most people believed that God created the world in seven days. Indeed, there are many that still believe this theory to be true. But on one of his many travels, Darwin had discovered evidence from fossils in rock, suggesting that there were animals that had lived many millions of years ago. On his return to Britain, he tried to discover how species evolved; the outcome of which was his theory of evolution, occurring by the process of natural selection.

Hormones Discovery of 1903

Physicists William Bayliss and Ernest Starling, introduced the term hormone in a lecture to the Royal College of physicians in 1905. Bayliss and Starling demonstrated that if acid was instilled into the duodenum, then it stimulated secretion in the pancreas. This was the first time that conclusive evidence of this phenomenon had been shown to other scientists.

Red blood cells 1663

Scientist Robert Hooke discovered the scientific theory of cells and the properties of cells. Through his cell research, Hooke was the first person to discover red blood cells, following huge advances in the world of microscope optics and magnification.

Stem cells first practical use 2008

Surgeon and physicist Martin Burchall led a team that created a new windpipe, grown by using stem cells, for a 30-year-old Columbian - Claudia Castillo. This was the world's first transplant organ using a patient's own stem cells to grow and replace an organ.

Test tube baby 1977

Physicists Patrick Steptoe and Professor Robert Edwards pioneered the technique of vitro-fertilisation. The result of the first successful operation was a baby girl called Louise Brown.

Monoclonal Antibodies 1986

Antibodies for experimentation and medical use, were previously derived from mice, making them difficult to use on humans due to the human immune system triggering anti-mice responses. It took the inventive approach of British Biochemist, Sir Gregory Winter to pioneer systems to fully humanise antibodies for therapeutic use.

"One general law, leading to the advancement of all organic beings, namely, multiply, vary, let the strongest live and the weakest die."

Charles Darwin

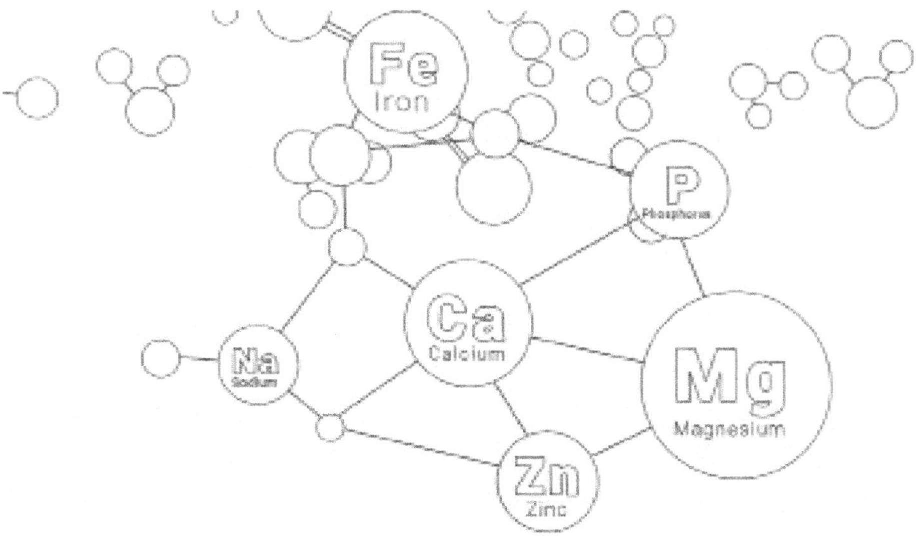

Chemistry

Argon Discovery and proof 1894

John William Strutt and William Ramsay were physicists who discovered the inert gas, argon, an achievement for which Strutt earned the Nobel Prize for physics in 1904. Argon is used in older style filament lamps to prevent oxidation, thus prolonging its life.

Benzene First isolation of 1825

Benzene, a natural hydrocarbon and an element of crude oil, was isolated from this liquid by scientist Michael Faraday in 1825 for the first time. Faraday was examining a viscous deposit that was a by-product of 'portable gas', made by plunging fish or whale oil into a white-hot furnace. This process created the gas which was condensed and then stored into portable containers for later use, to power gas lamps in public and private residences and commercial premises, such as shop displays. The oily liquid that was produced when this gas was compressed fascinated Faraday.

He successfully reported in a paper to the Royal Society, his successful separation of a new compound of hydrogen and carbon, 'which I may by anticipation distinguish as bi-carburet of hydrogen'. Following Faradays discovery, benzene was being manufactured on an industrial scale by the mid 1800s from coal tar, a by-product of coke manufacturing. It was used mainly as a degreasing solvent, but later in the 1900s was used in all manner of products such as, soaps and aftershave lotions.

Boron The isolation of 1808

Humphrey Davy was a chemist who isolated boron through the combining of potassium (K) and boric acid (H3BO3). It is noted that in the same year and by the same method, French chemists Joseph-Louis Gay-Lussac and Louis-Jaques Thénard also isolated boron.

Chemical electrolysis 1807

Sir Humphrey Davy was a physician and electrochemist who started experimenting with electrolysis by using a battery to prepare elements and split compounds. He discovered new metals such as potassium and sodium and electrolysed molten salts in 1807, and in the following year he revealed magnesium, barium, strontium and calcium. Davy's research into electrolysis and compound separation led to the new field of electrochemistry being founded.

Helium First observed on Sun's surface 1868

Sir Joseph Lockyer was an astronomer from Salcombe who discovered that the Suns atmosphere contained a previously unknown element called helium. It was named after the Greek God Helios, Greek for 'Sun' or 'the Sun God'. Lockyer identified helium, twenty-seven years before it was discovered as an element on earth.

Hydrogen Discovery of 1776

Henry Cavendish was a scientist, chemist and natural philosopher. Through experiment, his most notable discovery is that of hydrogen. He explained the concentration of 'inflammable air' that formed water when combusted. Antoine Lavoisier later replicated Cavendish's experiment and named the element hydrogen.

Nitrous oxide Discovery of 1772

Joseph Priestly, a clergyman and scientist discovered nitrous oxide, an odourless and colourless gas in 1772. Priestly produced nitrogen oxide by heating ammonium nitrate with iron filings, then siphoning the gas that was emitted, through water to detoxify, thus producing nitrous oxide. Priestley was also famous for isolating other key gases such as; sulphur dioxide, ammonia, carbon dioxide, carbon monoxide and oxygen.

Oxygen Discovery of 1774

Renowned scientist Joseph Priestly, made the discovery of oxygen 'O' in 1774. He produced oxygen by heating mercuric oxide (Hg0). At the time, Priestly called the gas 'dephlogisticated air', but the name of oxygen was actually created by Antoine Lavoisier who incorrectly thought that oxygen was essential to create all acids. Priestley was also famous for isolating other key gases such as; nitrous oxide, sulphur dioxide, ammonia, carbon dioxide and carbon monoxide.

Periodic table 1865

John Alexander Reina Newlands was a chemist who developed the Periodic table, now acknowledged as the benchmark in this field the world over. Newlands arranged the periodic table elements according to their atomic weights. He designed the now familiar periods and groups aligned in columns within the periodic table. Reina Newlands also devised the Law of Octaves.

Potassium First isolation of 1807

Sir Humphrey Davy was born in Cornwall. Davy discovered potassium in 1807, extracting it from caustic potash. Prior to the 1800s there was no distinction between potassium and sodium. The first metal to be isolated by electrolysis by Davy was potassium.

Sodium First isolation of 1807

Sir Humphrey Davy was the first person to isolate Sodium. Davy made this discovery through the electrolysis of NaOH, ultra-dry molten sodium hydroxide gathered at the cathode. Davy went on to isolate potassium using a comparable method the same year.

Thallium The discovery of 1861

William Crookes was a physicist and chemist who discovered thallium in residues of sulphuric acid production. Thallium is not found naturally on its own in nature. At the same time as Crookes' discovery, Claude-Auguste Lamy also made the same breakthrough, both using the newly created system of flame spectroscopy.

Viagra 1991

Andrew Bell, David Brown and Nicholas Terrett from Pfizer PLC, originally developed Viagra for heart patients suffering from angina, but after proving insignificant in treating heart problems, under the clinical guidance of Ian Osterloh, the team found that there was an unexpected side effect to this drug. Sildenafil as it was then known, could improve and sustain a man's penile erection. Pfizer immediately stopped research on this drug for heart treatment and it is now more widely known as Viagra, the world's first impotence treatment.

> "Living organisms are created by chemistry. We are huge packages of chemicals"
>
> David Christian

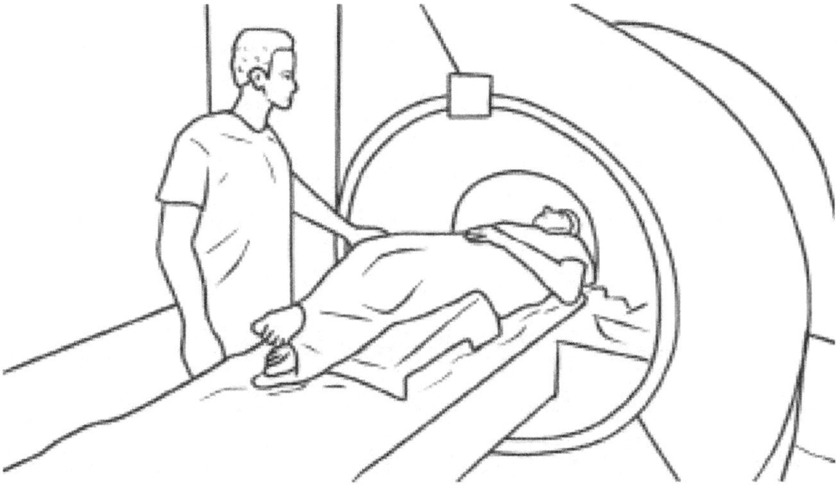

MRI Scanner

Medical innovation

Cataracts Artificial lens implant 1949

Ophthalmologist and surgeon Harold Ridley, pioneered cataract surgery. While working on the eyes of a fighter pilot who was the victim of a flying accident, he noted that the Perspex shards embedded in his patient's eyes were not rejected by the body's immune system. This 'Eureka' moment led to Ridley developing the first Perspex intraocular lens to be placed in the eye, enabling millions of people globally to see again, following cataract operations based on his pioneering work.

CAT or CT scanner 1971

Godfrey N Hounsfield was an electrical engineer. His development of the CAT scanner began when he constructed a computer that could process x-rays taken at all angles, thus creating an image of the object in segments. By applying this development to the medical field, he had created a process known as computed topography.

Hounsfield then built a trial head scanner and used a preserved human brain and then the fresh brain of a cow for sampling. He next tested the device on his own head. On October 1st 1971, CT Scanning was adopted by the medical world, primarily for head scans. In 1975, Hounsfield developed and built the first full body CAT scanner.

Cloning Theory of 1962

John B Gurdon is a scientist, recognised for his work in genetic cloning and stem cell research published in 1962. Gurdon was the first person to successfully clone an animal from a single cell and revolutionised the long development of stem cell technology into the 21st century.

Cloning First successful cloning 1996

Dr Ian Wilmutt is an embryologist from Warwickshire. He is most famous for leading his research team in the first successful cloning of an animal, delivering the now legendary 'Dolly the Sheep'. He was awarded an OBE in 1999 for his services to embryo development.

Colour blindness Discovery of 1794

John Dalton accidentally discovered colour blindness when he humiliated his family by buying his mother a set of bright scarlet stockings for her birthday. It was then that he realised that he was blind to this particular colour and his embarrassing experience had assisted him in understanding his own colour blindness because he had thought the stockings were blue. His brother also suffered the same problem. This was when Dalton realised there was a genetic link to colour blindness. His own colour blindness was the most common red-green variation and to this day it is still known as Daltonism.

Compound microscope 30 X magnification 1665

Robert Hooke was a prolific scientist, architect and inventor and he created the world's first compound microscope. In his publication on the device and its use, he called it the Micrographia. Hooke's microscope required the object to be placed on a glass slide and then natural light is diffused through the object, with the final adjustable lens close to the slide on the upper side. Hooke's design made the object appear detailed and clear against the bright background. The same design is still used in school and college laboratories the world over, only now the more sophisticated type uses a light beneath the slide for greater clarity and are commonly known as brightfield or illuminated compound microscopes now.

Contraceptive pill 1968

Chemist and physicist Herchel Smith developed the world's first cost effective and wholly synthetic contraceptive pill. He was born in Plymouth in 1925 and studied at both Oxford and Cambridge Universities. Smith later lectured at Manchester University, where he created and patented processes for making new steroids. He then moved to the USA and created a low-cost method to make a wholly synthetic hormone that prevented contraception called 'norgestrel'. This proved to be the beginning of low cost birth control. Later, 'Ovral' was developed and became available for general use in 1968. More than 800 patents were held by Smith for his research and production techniques, making him a multi-millionaire. He died in 2001 and left £45 Million to Britain's Cambridge University and several million to Harvard University in the USA.

Controlled clinical trials 1747

James Lind was a naval surgeon who performed what is now recognised as the world's first controlled clinical trial, during which he proved that citrus fruit prevented scurvy, a common disease of the day especially amongst mariners. Through the trials, he discovered that scurvy was caused by a lack of Vitamin C. Scurvy caused more deaths at sea in the 18th century, than enemy action. Lind also discovered that steam condensed from seawater could be drunk safely. It took many years before this process of water distillation was adopted, as there was no realistic method to boil up adequate quantities of salt water until the early 19th century.

Clinical thermometer 1887

Sir Thomas Allbutt was a physician and inventor of the world's first clinical thermometer. The end of the thermometer is put under the patient's tongue, under the armpit or into the rectum to reveal the body temperature of the patient. Allbutt's invention consisted of a thin glass tube with a bulb at one end that contained a liquid that expands uniformly with temperature. The tube has calibration marks along it with mercury or coloured alcohol used as the fluid. This method of determining body temperature in a clinical setting was used right up until the 2000s, after which more accurate digital thermometers became available for general clinical use.

Diabetes First synthetic insulin 1982

A team of British scientists from a subsidiary of Eli Lilly invented synthetic insulin. Prior to insulin being discovered, diabetes was a feared and fatal disease. Many doctors realised that sugar worsened diabetes and that a no-sugar diet was the most effective remedy, but not a cure. Then in 1921 a Canadian team of scientists discovered insulin and its effect on the human body. But it was not until the 1980s that a research team of British scientists from the Pharmaceutical Company, Eli Lilly, invented a synthetic substitute to replace bovine insulin along with its many side effects. The team inserted the gene for human insulin into a bacterial DNA and then used the bacteria as tiny factories to produce A and B chains of the protein individually. The second step involved a chemical process to combine them. It resulted in the world's first synthetic human insulin and it was called Humulin. Using Humulin as treatment for diabetes revolutionised not only the industry, but also the lives of millions of diabetic sufferers worldwide, because there are few side effects compared with animal insulin.

Dental chair Fully adjustable 1832

Inventor James Snell form London designed the world's first fully adjustable dental chair. Although the adjustments compared to today's chairs now seem negligible, it shaped future dental chair innovation in the years ahead.

Epilepsy Pioneer in treatment of 1858

Edward Henry Sieveking was a physicist, born in London. He discovered that the condition of epilepsy could be treated with the use of bromide and it became the first anti-convulsant in the world. He also published a significant theoretical advancement in epileptology, effectively the end of 'essential epilepsy'.

Hay fever discovery of 1819

More than one in four people suffer from hay fever in Britain alone and it's hard to believe that just two hundred years ago, nobody knew it even existed. But it took the determination of one man, himself a sufferer, to persuade the medical establishment to truly understand the scale of the phenomenon. John Bostock was a British doctor from Liverpool, whose main expertise was the study of bodily fluids, in particularly urine and bile. At the start of summer every year, Bostock suffered from sinus blockages, a runny nose and was generally lethargic for the next couple of months.

Then in 1819 Bostock decided to write a paper entitled: 'Case of a periodical affection of the eyes and chest'. It was a report effectively about Bostock's experience, detailing the symptoms and attempts of trying to allay the side effects by different treatments. These included, taking a cold bath, self-induced vomiting and taking opium, but nothing appeared to work. Determined to find a cure, Bostock expanded his research looking for and taking data from fellow sufferers.

By 1828 Bostock had located twenty-eight victims of hay fever and it was in a second publication that he named the condition as 'Catarrhus Aestivus' or Summer Catarrh. After this exhausting work, he decided to retreat to the seaside one summer and noted that he was almost rid of this 'summer catarrh'. Yet on his return to the countryside he noted that again his symptoms returned. It was at this point that he realised the link between summer flora and fauna that was in short supply at the seaside, was the most probable cause of the condition. This was confirmed by further research.

Hearing aid 1935

Edwin Stevens from Monmouth, invented the world's first truly portable hearing aid. Steven's device was the first 'wearable' hearing aid weighing in at a fairly heavy two and a half pounds. His device consisted of a base unit to fit over a belt, round the neck or in a pocket with acoustic amplifier and front facing microphone, and a wired earpiece to listen in. What made the unit so large and heavy compared to today's hearing aids is that the solid-state electronics used for the amplifier were in their infancy and battery technology and size had only just started to develop smaller more compact units, but still heavy and large by today's standard. He later developed smaller hearing aids that could fit completely around the ear as technology allowed smaller circuits and batteries to be created.

Heart lung & liver transplant 1986

Physicists and surgeons Sir Roy Calne & Professor John Wallwork performed the world's first triple organ transplant on Davina Thompson at Papworth Hospital, Cambridge.

Hip replacements 1961

Sir John Charnley was the pioneer of the world's first hip replacements. British surgeon and physicist John Charley created the world's first hip replacement in 1961 using a femoral stem and ball made of steel, with its hip socket made from Teflon. Several improvements have been made over the years, but more than 100,000 hip replacements are carried out now in Britain alone each year.

Hypodermic syringe 1853

The syringe in its most basic form had been around since ancient times, but it was Alexander Wood's innovative design that for the first time allowed drugs to be administered intravenously without the need to make an incision first in the patient's skin. Legend has it that his inspiration was when a bumblebee stung him. The Hypodermic syringe became a revolution in the world of anaesthetics globally.

NHS National Health Service 1948

The NHS was developed and launched on a philosophy that quality healthcare should be accessible to everybody, regardless of wealth. The NHS gave people in Britain, for the first time anywhere in the world, free Health care that provided on the basis of citizenship rather than ability to pay fees or buy insurance. The person behind this revolutionary way of thinking was British politician Aneurin Bevan who was instrumental in starting this great institution on 5th July 1948. The NHS remains a blueprint for many other countries around the world.

Parkinson's disease Discovery of 1817

James Parkinson was a scientist who notably published a paper in 1817 on 'Shaking Palsy', establishing Parkinson's disease as an internationally recognised medical condition.

Portable defibrillator 1965

Dr James Pantridge was a physician, cardiologist and professor from Hillsborough, who is famous for revolutionising emergency medical aid and improving paramedic services, through his development of the portable defibrillator. This machine treats life threatening cardiac attacks, by delivering a pulse of current to the heart, controlled by the operator. This invention is responsible for saving the lives of many thousands of people world-wide due to its portability and local availability, without the need to visit hospital.

Medicine

Acetylcholine 1914

Henry Hallet Dale was an inventor who discovered acetylcholine, the chemical compound that is a neurotransmitter. It was noted for its actions on heart tissue in particular and was confirmed as a neurotransmitter later by Otto Loewi. Both Scientists received the 1936 Nobel Peace Prize in physiology or medicine for their pioneering work in this field. Acetylcholine functions both in the central nervous system (CNS) and in the peripheral nervous system (PNS).

Anaesthetics Nitrous oxide 1799

Nitrous oxide was first discovered by Sir Humphrey Davy and is recognised as the world's first commercial and practical anaesthetic, now better known as laughing gas, still used to this day worldwide in many hospitals as a reliable, safe form of anaesthesia with very few side effects.

Anaesthetics Chloroform 1847

Obstetrician Sir James Young Simpson, was a hugely important figure in the chronological history of medicine. He discovered the anaesthetic properties found in chloroform and introduced it successfully for general medical practice during an experiment with some of his friends, where he soon discovered that it could be used to put someone to sleep. It was later discovered that chloroform was highly toxic to the liver.

Anaesthetics Halothane 1951

Discovered by chemist, Charles Suckling while working at ICI (Imperial Chemical Industries). He first synthesized halothane, then an extremely volatile anaesthetic in 1951, whilst working in his ICI laboratory in Widnes Lancashire. Following Suckling's first investigations on the anaesthetic properties of halothane by experimenting on houseflies and mealworms, he then passed it on to a renowned pharmacologist, Jaume Raventos for further evaluation on other animals.

After establishing its medical properties and possible application, Reventos then provided his findings to a prominent Manchester anaesthetist, Michael Johnstone, who recognised its huge advantages over conventional anaesthetics, establishing its first clinical trial in 1956. This has been documented and recognised as one of the first instances of modern drug design in the world.

Antiseptic surgery 1865

Joseph Lister was a surgeon who introduced antiseptics to surgery and subsequently employed them to increase the success rates in theatre. Open wounds offered easy access for germs to enter the body. Lister read that carbolic acid had been used to treat raw sewage and to kill cattle parasites. This provided him with the notion that it could also prevent infection of wounds. He then started to use Carbolic acid to clean the wounds of his patients and at the same time soak the dressings in the same antiseptic fluid. Over time, Listers death rate in surgery decreased from 45% to below 15% after introducing his new antiseptic treatment.

Aspirin Discovery of the active ingredient 1763

Edward Stone was a Church of England clergyman who was walking through a meadow one day near his home at Chipping Norton, suffering from aches and pains. He was alleged to randomly rest by a willow tree and broke off a small piece of its bark and started nibbling at it. He noted its remarkably bitter taste and knowing that the bark of the Peruvian tree cinchona produced quinine with a similar bitter taste (used to treat malaria) he deduced that willow might also have similar therapeutic qualities.

Following several experiments of drying willow bark and pounding it to a fine powder, he gave it to about 50 different people to test.

Stone concluded that it was found to be very effective in curing fevers and relatively mild disorders such as headaches. He discovered in the process of testing that the active ingredient was salicylic acid. He sent a letter confirming his discovery to the president of the Royal Society. The letter survives to this day.

Aspirin Effectiveness 1982

John Vane was a pharmacologist who was a key figure in understanding how pain and anti-inflammatory effects were produced by the drug Aspirin. His research led to new treatments for blood vessel and heart disease, including introducing ACE inhibitors. Vane was co-awarded the Nobel Prize in physiology and medicine in 1982 with Bengt I. Samuelsson and Sune K. Bergstrom for their findings relating to prostaglandins and biologically active substances.

Beta blockers Propranolol 1964

The science of cardiology takes in some of the most respected names in medicine. One such person was physicist Sir James Black who contributed hugely in this specialist field, as a scientist and a physician. His work in cardiology and invention in 1964 of Propranolol was a beta-adrenergic receptor antagonist and it revolutionised the treatment and stability of the heart condition angina pectoris. It is regarded as one of the most significant contributions to pharmacology and clinical medicine of the 20th Century. In 1988 he was awarded the Nobel Prize in medicine.

Blood circulation Discovery of 1628

William Harvey was a physician and the first person to demonstrate and prove that blood circulates around the body. Previous to this discovery, it was commonly thought that blood flowed freely throughout the body similar to a bag full of liquid, rather than being transported via a series of arteries and veins.

Blood pressure measurement 1733

Chemist and botanist Stephen Hales, famously took blood pressure measurements in several species of animal by inserting very fine tubes into veins and arteries, then measuring the height that the column of blood rose. This was to be the world's first practical blood pressure test.

Blood transfusions Human to human 1818

Physician, innovator and pioneer James Blundell performed the world's first human blood transfusion. It was in the pursuit of safe and effective blood transfusion that Blundell was most interested and motivated. He first experimented with whole blood transfusion between dogs before successfully carrying out experiments from human to human using syringes, taking care to avoid air intake into the veins. Blundell also started to perfect a device that now resembles a modern-day drip to introduce appropriate drugs in a 'controlled manner'.

Blood transfusion service 1921

Physician Dr Percy Oliver started the world's first voluntary blood transfusion service. Under Oliver's supervision, British Red Cross members elected to give blood at King's College Hospital London; the start of the world's first Voluntary Blood transfusion service.

Brain 'locator GPS' Discovery 2014

A British based research team led by Professor John O'Keefe with May-Britt Moser and Evard Moser shared the Nobel Prize for physiology or medicine in 2014 for discovering and explaining the brain's 'GPS' system. The team based at University College London, discovered how the human brain knows where we are and how to plot a course from one point to another. It is hoped that this discovery will help solve some of the unanswered questions relating to Alzheimer's disease, where patients cannot recognise their surroundings. Professor O'Keefe discovered the first part of the brains positioning system as far back as 1971. Then in 2005, husband and wife team May-Britt and Edvard Moser found a different area of the brain that performs similar to a maritime chart with grid like sections. This they found helps humans to navigate and judge distance. Moreover, according to Dr Colin Lever from the University of Durham (one of Professor O'Keefe's team), he found that the pioneering work and discovery, supports the theory that humans possess an autobiographical memory.

Cholera Discovery that it was water bound 1854

John Snow was a physician, expert in medical hygiene and anaesthesia. He traced the source of a major cholera outbreak in London in 1854. His findings pointed to the water borne diseases carried within London's poor water and waste systems and this led to similar recommendations for change in other large British cities. His research ultimately led to important improvements to general public health worldwide.

DNA Discovery 1953

Francis Crick (GB), James Watson (USA), Maurice Wilkins (NZ) won the Nobel peace prize for the discovery of the structure of DNA in 1962, although they first made the discovery and published their findings on the structure of Deoxyribonucleic Acid (DNA) in April 1953. Rosalind Franklin (GB) was also working with Wilkins at King's College London and were both using the technique of X-Ray diffraction to study the DNA. Shortly after, they used their findings to publish their own work on their discovery, a molecular structure of DNA based on all previous research know characteristics, and the now famous double helix. They used this model to explain how DNA replicates and its relationship to hereditary information coded on it.

Female Doctor 1857

Physician, doctor Elizabeth Blackwell was the world's first woman to receive a medical degree in the USA and become the first woman to be registered on the British Medical Register. Blackwell later broke new ground in encouraging the education of women in medicine both in the USA and Great Britain.

Gray's anatomy 1858

Anatomist Henry Gray, studied the development of the spleen and endocrine glands in 1855. It was through this research that he made a proposal to his colleague Henry Vandyke Carter, with an idea to develop and publish an accessible textbook that would be suitable for medical scholars. Vandyke Carter was agreeable to this project and so the two set out on their crusade of dissecting unclaimed bodies from mortuaries and workhouses for the next two years, to gather data and drawings that formed the basis for their book. The book was completed and published in 1958 in Britain and the following year in the USA. It was called Gray's Anatomy. It was the first published anatomical textbook in the world.

Hypnosis 1841

James Braid is recognised as the first authentic hypnotherapist. Braid was a scientist and practicing surgeon who was the single most significant innovator in the field of hypnosis. Braid saw the benefit and treatment by hypnosis as complementary to primary medicine, rather than an alternative. In his experimental stages, Braid revealed that hypnotism on a human specimen simply set off a chain of invisible sensual events that merely required gentle support to achieve their aims and progress. The modern science of hypnosis certainly owes its name and functional principal to Braid's extensive research and development in this field.

Ibuprofen 1961

Ibuprofen is a drug now used the world over and belongs to the drug family known as anti-inflammatories and the world's most popular drug in that group. Chemist Stewart Adams and his team, working for drug Company Boots PLC in Nottingham, produced this ground-breaking drug after several years of research and patented it in 1961. Ibuprofen is now one of the most widely used drugs in the world.

IVF 1978

Patrick Steptoe was a pioneer of fertility treatment, gynaecologist and obstetrician & Robert Edwards was a biologist and physiologist. Together they broke new ground in the field of in vitro Fertilization. The first test tube baby in the world was Louise Joy Brown and she was born on July 25th ,1978. Steptoe died in 1988, but Edwards was awarded the Nobel Prize in physiology or medicine in 2010. Unfortunately, Steptoe was not considered for the Nobel Prize because it is not awarded posthumously.

MRI Magnetic Resonance Imaging 1973

Sir Peter Mansfield, a physicist from London, is credited as demonstrating how radio signals from MRI can be mathematically evaluated. Mansfield's research enabled a system to be devised in 1973 that could extrapolate the signals of an MRI device and transpose them into a practical image. Mansfield's work is also acknowledged to have made rapid practical Magnetic Resonance Imaging possible by developing the 'MRI Protocol' known as Echo Planar Imaging.

MRI Machine 1980

The world's first full body MRI scanner was built and trialled by a team led by professor John Mallard including professor James

Hutchinson and doctor Bill Edelstein. On August 28th, 1980 Mallard's Magnetic Resonance machine based on Sir Peter Mansfield's MRI Protocol, obtained the first clinically useful image of a patient's internal organs. The first patient it was used on following trials was a man from Fraserburgh revealing a cancerous stomach tumour. The MRI scan revealed that the patient also had secondary cancer in his bones and an abnormal liver.

Penicillin 1928

Physicist and inventor Sir Alexander Fleming discovered how penicillium notaturn reacted on bacteria and made the forerunner to modern antibiotics, as we know them today.

Penicillin Mass production 1945

Biochemist Dorothy Crowfoot-Hodgkin discovered the chemical structure of penicillin and through her pioneering research work, it is now the worlds most widely used anti-biotic. Later that same year, a team of scientists at the world-famous Oxford University, research scientists developed a way to mass-produce the drug. The ground-breaking team included British scientist Norman Heatley, Australian scientist Howard Florey and German scientist Ernst Boris Chain. Chain and Florey shared the 1945 Nobel Prize in Medicine with Sir Alexander Fleming for their work.

Malaria 1897

Ronald Ross was an Army Surgeon who established the cause of Malaria by discovering the malarial parasite in the infected gastrointestinal tract of mosquitoes. Ross's documented discovery paved the way for other scientists to find a cure to combat the fatal disease.

Smoking & lung cancer link 1950

Physicists Sir Richard Doll and Sir Austin Bradford Hill first proved that smoking causes lung cancer. At the time of their discovery, more than eighty percent of the British population smoked cigarettes, causing a shocking rise in the number of smoking related deaths from Lung Cancer. Bradford-Hill and Doll were asked to investigate this alarming phenomenon. At first it was thought that the rise in lung cancer directly correlated with the huge road-building programme, the fumes from Tarmac-laying and increased car fumes, as ownership was rapidly increasing. But then the breakthrough came; in a sample of seven hundred lung cancer patients, only two were non-smokers. The study of lung cancer patients continued, and it provided conclusive evidence that smoking caused the disease.

Steri-spray Sterilises Water 2008

Ian Helmore was a plumber whose job it was to sterilise water tanks to prevent the breeding of legionella. He discovered that bacteria might remain in the last few inches of pipe work, so while experimenting to find a solution, he discovered that installing an Ultra-Violet lamp inside a spray head or tap would deal with the issue. The Steri-spray instantly sterilises water. This simple system is now used in the NHS and hospitals the world over.

Surgical forceps 1727

Scientist Stephen Hales is attributed to the invention of surgical forceps. It was and still is a hand-held device for holding or grasping objects, often used when fingers or hands would be too large to hold small items.

Typhoid Discovery and prevention 1838

William Budd was a medical doctor whose investigative work on typhoid fever had huge significance in the years to come. Budd served his early years as a trainee doctor with French surgeon and pathologist Pierre Louis, who had a special interest in an illness of the gut called putrid fever. Budd was also influenced by the work of another French doctor, Bretonneau who also worked on treating a similar fever. It was Budd's experience in France that encouraged him to look at a similar disease in Britain known as Typhoid Fever, due to similarities to typhus, though not connected.

Budd ultimately discovered the casual bacterium, salmonella typhi. He first published a paper on the subject in 1938 but made his discovery proper 42 years later in 1880. He concluded that rather than being a contagious disease, typhoid fever was in fact caused by poor sanitation and water hygiene and the subsequent ingestion of contaminated water and food. This led him to become an enthusiast in disinfection and the promotion of clean water and drainage systems.

Typhoid First vaccine 1896

Medical doctor, Almroth Edward Wright was a supporter of William Budd's discovery that typhoid fever is non-contagious and its causes thereof. Wright set about experimenting with an anti-dote to the fever and in the course of his research, developed a successful inactivated whole-cell typhoid vaccine launched in 1896.

However, the vaccine was found to have some serious side effects and its use was stopped a few years later. To date, although several vaccines are on trial, not one has been sanctioned for commercial use since Wright's invention, with doctors having to use strong antibiotics, rather than administer immunisation without side effects, still the long-term goal.

Ultrasound imaging 1956

Physicist Ian Donald was the first person to introduce ultrasound into medicine for diagnostics in 1956. With the aid of a one-dimensional amplitude mode (A Mode) device, he was able to measure the diameter of the foetal head. Then, two years later, Donald displayed an ultrasound image of a female genital tumour. Ultrasound scanning is safe and used throughout the world now to scan and display various stages of the unborn child.

Vaccination 1796

Edward Jenner was a medical doctor from Berkeley, Gloucestershire, who pioneered the small pox vaccination, and was known as 'the father of immunology'. Jenner noticed that milkmaids who suffered from the mild disease of cowpox, never ever contracted smallpox, one of the biggest killers of the time. So he famously placed pus taken from a cowpox cyst, into an incision on eight-year old James Phipps arm. Jenner was able to prove that Phipps was immune to smallpox, having been vaccinated with cowpox. He later carried out several more vaccinations on other young children including his own son, to prove his theory. The word vaccine was taken from the Latin word 'vacca' for cow.

> "The art of medicine consists in amusing the patient while nature cures the disease"
>
> Voltaire

Physics

Bragg's law 1912

Scientist and physicist, William Lawrence Bragg submitted his derivation of the reflection condition to the Cambridge Philosophical Society at a meeting in November 1912. His paper was published in 1913. The Bragg's law concept provides the condition for a plane wave to be diffracted by a collection of lattice planes. Its use in the development of electro-magnetic radiation, particle waves, gamma rays and x-rays in relation to electrons and neutrons has been invaluable to modern day physics and its progress.

Conservation of energy Law of 1843

James Joules was a brewer, who at the age of 22 revealed that heat generated in a coil of wire is directly proportional to the square of the current running through it. The effect is now called 'Joule Heating' and in recognition of Joules' importance in demonstrating energy conservation, the SI unit and the scientific term known as the 'Joule' is named after him.

Dalton's law of partial pressures 1801

Eminent scientist John Dalton, discovered the relationship of the individual pressures of each gas in a mixture of gas. Dalton's Law states that the combined pressure of a mixture of gases is equal to the partial pressures of the component gases.

Diamagnetism 1845

Prolific scientist and inventor Michael Faraday discovered that several substances show a faint repulsion when introduced to a magnetic field and he named this phenomenon diamagnetism.

Faraday also observed that the he could affect a ray of light in the direction that it was moving by introducing a magnetic field into its path. He noted that he had 'succeeded in illuminating a magnetic curve'.

Electric transformer 1821

Electric generator 1831

Electric motor 1831

There is good reason why all three pioneering inventions above are grouped together. Michael Faraday is best known for his discoveries of the laws of electrolysis and electromagnetic induction. Faraday's greatest breakthrough was his invention of the electric motor in 1831.

However, Faraday's first development was the electric transformer in 1821. Then 10 years later, he created an ingenious device based on a ring and discovered when an iron core was rotated in the centre of the ring of wound wire; it generated a potential across the two wires. This was the world's first generator. Further experiments followed on the same device, when he applied a potential or Voltage to the two wires on the generator, then the iron core would rotate. This was the world's first electric motor. All three related inventions were absolutely pivotal in industrial progress and are probably the three most important inventions of the 19th Century.

Faraday went on to develop many other significant discoveries, such as benzene, the 'Faraday effect' and the legendary 'Faraday cage' and finally he had the unit of electrical capacitance; the 'Farad', named after him. Faraday refused the title of a Knighthood and presidency of the Royal Society, instead devoting his working life entirely to scientific research.

Electro magnet 1823

William Sturgeon was an inventor and self-taught physicist and mathematician, who in 1823 demonstrated his first simple electromagnet. He revealed its power by lifting a nine-pound weight with just a seven-ounce rod of iron, wrapped with eighteen turns of bare wire that was energised with current from a single cell.

Electron 1897

Although many scientists before 1897 had suggested that the atom was constructed from a more elementary unit, it was British physicist and inventor JJ Thomson, who was the first to propose that the elementary unit was more than one thousand times smaller than an atom. Thomson advocated that there was a subatomic particle, now known as the electron.

Faraday cage 1836

Michael Faraday, a prolific scientist, electrical engineer and inventor, developed a shield made of conducting material, normally mesh, to block external electrical fields, both static and non-static, therefore providing a constant Voltage on all sides of the enclosure. The Faraday cage is used in situations where electronic equipment needs to be protected from static or non-static discharges.

Fuel cell 1842

Physical scientist and judge William R. Grove, discussed the universal theory of the conservation of energy, but more importantly, pioneered the development of fuel cell technology. He developed a device called the 'gas Voltaic battery', that was to become the world's first working fuel cell that produced electricity. Grove's device produced a current by mixing oxygen and hydrogen. He developed and demonstrated, that the cell could prove that steam can be converted into hydrogen and oxygen and the same process could then be reversed. Grove was the first person to show thermal disassociation of molecules into their component atoms.

Graphene 2004

Scientists recognised the existence of graphene and its remarkable physical characteristics and potential applications for the future. But, the big breakthrough came when two British based researchers at the University of Manchester, Professor Andre Geim and Professor Kostya Novoselov, isolated graphene in 2004. Even though physicists were aware that one atom thick, 2D crystal graphene existed, up to this point, nobody had worked out how to extract it from Graphite. Forecast to transform the world we live in, it is completely transparent, yet conducts both heat and electricity and is billed as the world's first 2D material. Graphene is a technology that has the potential to disrupt medicine, travel electric vehicle technology and a host of other industries. Both Scientists received the Nobel prize for Physics for their ground-breaking work.

Gravity The law of 1665

Legend has it, that physicist Isaac Newton was one day sitting under an apple tree, when an apple fell on to his head. It was at this moment that he started to think about the law of gravity. Although this is almost definitely not the case, the story acts as a perfect anecdote to describe what actually happened. What is sure is that Newton started looking into why things fall or travel and the cause and effect of such a phenomenon.

Newton deduced that if a matter is accelerated through falling from a tree, then because its speed changes from zero on the tree, it moves towards the ground, therefore there must be a force in action and this force must be due to gravity. Newton reached the conclusion that any two objects in the universe will exert gravitational attraction towards each other and that the force will have a common form.

Hydroelectricity 1880

William Armstrong was an inventor and engineer from Newcastle and is credited with designing and installing the world's first house in the world, to be powered by hydro-electricity. Hydro-electricity is now the most widely used form of power in the world.

Isotopes First concept of 1912

Economist and chemist Frederick Soddy, proposed that the same elements exist in different forms. He cited that nuclei having the same number of protons could have different numbers of neutrons. So, his theory of isotopes explained that different elements can be indistinguishable chemically, but at the same time they can have completely different atomic characteristics and weights. His theory led to a huge breakthrough in 1920 when he displayed the significance of isotopes, in calculating geological age. This led to the huge breakthrough of carbon dating.

Murdoch 1792

William Murdoch was an inventor and engineer. After working as an engineer building steam engines for the Cornish company Boulton & Watt, he then moved to the heavily industrialised City of Birmingham. At the start of the 1790s, he embarked on a scheme to find a way to turn gas efficiently into a white light for illumination, rather than conventionally using oil and tallow, popular at the time.

It was in 1792 that he started experimenting with gas extracted from coal and other combustible materials. Murdoch's first experiments were very crude, with gas produced from burning coal in a small steel container, through which a tube directed the gas by forcing it up a rifle barrel and the escaping gas was then lit, producing light. Over the following years, Murdoch carried out a series of experiments based on the quantity and quality of the gasses he found in different materials and looked into finding the best ways to store and transport these hazardous gasses. This led Murdoch to refine and build the world's first portable gas lamp in 1796.

Jet propulsion Theory 1686

Scientist Sir Isaac Newton developed the principal of jet propulsion. He also forecast that one day people would be able to travel at speeds up to '50mph' in a jet powered vehicle, based on his 'third law of motion' theory 'to every action there is an equal reaction'. But it wasn't until 1774 that a man called Gravesande developed a steam powered Jet Car, when a boiler mounted on the cars chassis pushed the vehicle along by emitting a powerful jet of steam.

Joule Thermodynamics 1843

Scientist James Prescott Joule discovered the relationship between heat and mechanical movement. Joule's discovery then led to the law of conservation of energy, in turn leading to the first law of thermodynamics. Joule also worked with Lord Kelvin and helped develop the absolute scale of temperature. Joule further observed the relationship between the current flowing through a resistor and the dissipated heat. This is now called Joules first law.

Kaleidoscope 1816

Sir David Brewster is renowned for his invention and development of the kaleidoscope. A tube -shaped toy, containing loose pieces of coloured glass or plastic and mirrors that reflect changing patterns as the tube rotates. He was a mathematician, physicist, inventor, writer, science historian, University principal and astronomer. He is most famous for his involvement in the field of optics. Brewster discovered the photo-elastic effect, giving birth to the study of optical mineralogy. But it was his work and development on the Kaleidoscope, and the lighthouse illuminator that gave rise to his greatness in the field of optics, during the 19th century.

LED Theory of 1907

The first report of a solid-state light emitting diode being made and tested, was in 1907 by British engineer HJ Round of English company Marconi Labs. He was a personal assistant to radio pioneer Guglielmo Marconi. Round used a crystal of silicon carbide and a cat's whisker detector. He was educated for much of his schooling at Cheltenham Grammar School for boys, and later attended the Royal College of Science, where he was awarded a first-class honours degree. He was successful in submitting and receiving 117 patents.

Light bulb Electric incandescent 1800

Eminent scientist Sir Humphrey Davy, demonstrated the world's first incandescent light bulb in 1800, by passing electricity through a strip of carbon. The carbon glowed brightly, producing an arc light, but soon burnt out after a few seconds. Davy later experimented with platinum and used a vacuum pump to extract air from the lamp, before it was sealed to ensure the filament lasted longer.

Light bulb 1880

Joseph Swan created the world's first practical light bulb, beating Thomas Edison to the patent by producing a reliable alternative to the gas lamp. But for both men, their creations had extremely short life spans of just a few hours. Eventually, both men decided to pool their resources and co-develop their inventions, by forming the Edison Swan Company.

Light switch Quick break technology 1884

Inventor John Henry Holmes from Newcastle-upon-Tyne, invented the world's first quick-break light switch in 1884. Holmes solved the problem of switches arcing each time the circuit was opened or closed and solved the solution, by ensuring that the contacts opened or closed rapidly each time the switch was operated. Less arcing meant longer life contacts and the Holmes switching principle is still used internationally in all light switches.

Light Nature of electromagnetic theory 1873

Professor and scientist James Clerk Maxwell, experimented with magnetism and electricity, concentrating on electromagnetic theory and light, discovering in the process that electricity travels at the speed of light. He also recognised that magnetism and electricity are one of the same thing, known as electromagnetism, and in the process of his research predicted the existence of radio waves.

LCD Liquid Crystal Display 1970

It is quite possible that you are reading this article on a device, made possible only because of British inventor, George Gray's creation of the LCD. Gray developed the molecules that enabled LCD's to function. In 1970, Grays work on LCD's was first published, initiating a multibillion-pound industry, making today's profusion of LCD flat screen gadgets achievable.

Lithium-ion Battery 1976

The world's most popular rechargeable battery, started life with a team of researchers and scientists at Oxford University's inorganic chemistry laboratory. In the mid-seventies, there was an upsurge in development by companies and scientists, searching for more compact, lighter and longer lasting power storage solutions, that had hitherto been thwarted by a succession of setbacks.

Led by John Goodenough, his research group included, Dr Phil Wiseman, Dr Koichi Mizushima and Dr Phil Jones, the team looked at various ways of employing lithium cobalt oxide. After much experimentation (and several fires in the lab), the group published in the Materials and Research Bulletin in 1980. The first commercially manufactured Lithium-ion battery was subsequently launched by Sony 10 years later.

The lithium-ion battery is now ubiquitous and is powering the huge growth of electric transportation globally, while the world seeks to reduce carbon emissions in every part of our daily lives.

Magnetic Earth 1600

William Gilbert was a Cambridge scholar, natural philosopher and physicist from Colchester, credited by many as being 'the father of electricity and magnetism'. Britain was a dominant seafaring nation during Gilberts lifetime and mariners relied a great deal on the magnetic compass for Navigation. There was much mystery about where magnetism came from up to this point, with some eminent sailors such as Columbus thinking that the pole star affected the compass needle. This prompted Gilbert to carry out experiments in an attempt to clarify the phenomenon of magnetism once and for all. During his experiments, he noticed that magnetic forces frequently created circular motions, pointing to a link between this phenomenon and the Earth's rotation. This provided Gilbert with a theoretical basis for further research into the science of magnetism.

Magnifying glass 1250

Roger Bacon was a philosopher and inventor from Somerset. He graduated at Oxford University at the age of 13 and later became a master at his college. In the course of his many experiments, he was the first person to invent and document the magnifying glass, whilst researching the eye, reflected vision, refraction lenses and mirrors. Bacons invention happened long before the patent system was established and was therefore freely copied and enjoyed by many for years to come.

Mass spectrograph 1919

Scientist Sir William Aston, is credited with building the world's first fully functional mass spectrograph in 1919. Aston was continuing his work with renowned scientist J.J. Thomson and during Aston's experiments with the fully built device, he was able identify isotopes in chlorine, krypton and bromine, providing positive evidence that these natural elements are made up of a mixture of isotopes. Aston's use of electromagnetic concentration in his Mass Spectrograph, allowed him to quickly detect at least 212 out of the 287 naturally occurring isotopes. The mass spectrograph is still in use today, allowing crime laboratories, water authorities and science researchers worldwide, when there is a need to identify a substance that may contain many other different elements.

Microwave 1864

James Clerk Maxwell was a brilliant inventor who discovered the existence of microwaves, by the use of and proof through his famous paper of 1864, using proven equations and calculations. Based on his research, 24 years later, Heinrich Hertz proved Maxwell's claims by constructing a device that detected microwave radiation. Many scientists believe that Microwave discovery is possibly one of the greatest inventions of the 19th Century.

Motion the law of 1687

Isaac Newton was one of Britain's greatest mathematicians and scientists. During his latter years at University, Newton was publishing some of his ideas in a journal about his thoughts on motion, calling them the three laws of motion. First law: often called the law of inertia; an object at rest will continue at rest unless acted upon by an unbalanced force. All objects resist change in their state of motion. Second law: when force acts on a mass, then acceleration is produced. Greater force is needed to accelerate the mass, the greater the mass is. Third law: for every action, there is an equal and opposite reaction. This means that an object will get pushed back equally as hard as it is being pushed in the opposite direction.

Neutron 1932

James Chadwick was a scientist who in 1932 followed a field of research that led to the discovery of the neutron. He quickly went on to calculate its mass. He predicted that neutrons could develop into a major breakthrough in the fight against cancer. Chadwick was also involved in projects developing the atomic bomb during his war years.

Nuclear power generation station 1956

The world's first commercial nuclear power station was at Calder Hill, Cumbria, also known as Seascale. When it first opened it was a reason for great pride and celebration, signifying the beginning of the fledgling atomic age. The site was closed in 2003 and commenced decommissioning, a process that will take up to 100 years to complete. It was also the world's first Magnox reactor, so called because the fuel vessels were produced from magnesium alloy.

Radio first transmission and receiving 1856

Engineer David Edward Hughes, is documented as the first person in the world to transmit and receive radio waves. He designed and developed the synchronous type-printing telegraph. This was 40 years before Marconi (1896) was credited with sending a wireless telegraph and 10 years before Mahlon Loomis of Washington USA, sent signals by radio (1866).

Radio Theory of electromagnetism 1873

James Clerk Maxwell was a mathematician and scientist, who through his research in the fields of electricity and electro-magnetism pioneered the concept of the 'Electro-Magnetic field'. It was during this research that James Clerk Maxwell proved the existence of radio waves and this discovery generated notional connection, that light waves were also connected to this electrical phenomenon. Thereby discovering the existence of radio waves, paving the way for even greater discoveries.

Solar power First Solar cell patent 1888

Edward Weston was a chemist from Shropshire, who was acknowledged for his work in the fields of electrochemical cells, electro-plating and most significantly, the 'Weston Cell'. His solar cell patent was described as 'apparatus for utilising solar radiant energy' and lodged for patent approval in New Jersey during 1888. This early solar cell was very inefficient, at less than one per cent and it took more than fifty years before Russell Ohl, who patented the Silicon semiconductor solar cell in 1946, developed a serious alternative. Even solar cells made today still only operate at less than thirty per cent efficiency.

Spectrum Heterogeneity of light 1655

Prolific inventor, scientist and physicist, Sir Isaac Newton carried out a number of experiments between 1642 and 1655 indicating that he was the first person to understand the rainbow, by refracting white light through a prism, thus transforming it into its constituent colours of red, orange, yellow, green, blue and violet

Speed of light 1728

Clergyman and amateur astrologer James Bradley discovered the 'aberration of light'. In its own right this phenomenon is not that significant, but it encompassed and allowed Bradley to prove beyond reasonable doubt, that the planet Earth was travelling through space and proved that the universe did not revolve around Earth. Furthermore, through this discovery he was also able to calculate that light travelled at 670,616,629 mph, incredibly within only 1% of the accepted figure agreed by scientists today.

Telescope Reflecting 1668

Sir Isaac Newton, the famous scientist and inventor was a fellow at Cambridge University's Trinity College. He developed the idea of a reflecting telescope and made it reality in 1668. This was a huge step in the technology of telescopes, because it made astronomical observation far more precise, by using two concave mirrors inside the main tube, rather than lenses to produce a magnifying effect.

Wind-up radio 1991

Trevor Baylis was watching a TV programme filmed in Africa about AIDS. During the documentary, the presenter commented that a sure-fire way to stop the spread of the disease would be for all people to have access to good information on the radio. It was at this point that Baylis designed a radio that required no batteries, instead, taking its power from a wind-up internal spring powered by a hand crank. He demonstrated a model to Nelson Mandela and ever since that meeting; his radios have been distributed all over Africa.

Zero Devised scale to show absolute zero 1848

It was not until 1848 that William Thompson or Lord Kelvin developed Robert Boyles original theory on absolute zero. Lord Kelvin decided that it would be helpful to be able to characterise extremely low temperatures precisely. The breakthrough came when he noted that molecules stopped moving at absolute zero. His rules were known as the Kelvin scale, where absolute zero is minus 273.15 degrees on the Celsius scale. His work is vitally important in the field of superconductivity, where superconductors are extremely efficient at conducting electricity only at low temperatures; discovered only after Kelvin's death.

> "Physics tells us observations can't be predicted absolutely. Rather, there's a range of possible observations each with a different probability"
>
> Robert Lanza

Rugby

Sport

Introduction to modern sports

The majority of modern sports can be traced back in their creation to Great Britain, not just in standardisation and rules, but also their origins and development. Most significant of which are tennis, cricket, rugby, football, squash, hockey, golf, snooker, billiards, darts and table tennis. More sports can also trace their roots back to British sports, such as: American football (rugby) and baseball (rounder's). It's clear that not every ball game was originally created in Great Britain, but what did happen was the establishment of British standardisation in both rules and structure for most of the world's most competitive sports in the world today.

Badminton Racquet sport 1873

Badminton was first played in the form that we know today, at a party held by the Duke of Beaufort at Badminton House in Gloucestershire. Its origin in its most basic form probably dates back more than 2000 years. However, in the 1600s battledore and shuttlecock was an upper-class pastime in England that spread to many European countries. Badminton was played by two people hitting a shuttlecock back and forth, without allowing it to hit the ground.

In India, a net was introduced to the game by British officers in the 1800s, to add more of a challenge to their pastime game; then called poon. When officers brought this game back to England, it was introduced to the Duke of Beaufort during 1873 at his home in the stately manor of 'Badminton' in Gloucestershire, where the game went on to spread and become more popular.

In 1898, the first open tournament, now known as Badminton, was held at Guilford in the world's first 'all England' championships. During the 1930s the USA, Denmark and Canada all became passionate followers of the game. In 1934, the International Badminton Federation (IBF) was formed. Founding members were, England, Wales, Scotland, Ireland, France, New Zealand, Holland, Canada and Denmark. India joined as an affiliate in 1936.

Baseball 1755

Documented by William Bray's diary entry. It is now confirmed that local historians from Surrey have found conclusive evidence that baseball was played in Britain more than 20 years before American independence. The diary belonging to local lawyer and diarist William Bray has now been attributed as the first description and mention of the game of baseball. It describes the game being played in Guilford by men and women in March 1755 on Easter Monday.

Basketball 1891

The British game of basketball is a sport invented by British doctor, James Naismith and first played in 1891. In the UK it later evolved into the game of netball, now predominantly played by women, except in the USA where basketball is now principally played by men's teams.

Billiards 19th Century

The game of billiards is generally played by two players or teams and originates in Great Britain, spreading rapidly to the colonies during the 19th Century. There are two cue balls in billiards, both white, but one is marked with a black dot and a red object ball is used. The game is played on a table of similar size and dimensions as a snooker table. There is a different cue ball assigned to each player or team and points are scored for cannoning and pocketing the balls in any one of the six pockets around the edge of the table.

Bowls The sport 1299

The British sport of bowls has been traced back to the 13th Century. The world's oldest bowling green still played on is in Southampton. Records reveal that it has been functioning in its current form since 1299. The game is now played globally.

Boxing Queensbury rules - current rules 1865

John Chambers was the Marquis of Queensbury, who publicly sanctioned the world standard code of boxing, written by John Graham Chambers. The code is still the basis for professional boxing rules used to this day. The rules were the first to mention the compulsory wearing of gloves and designed to be used for both amateur and professional boxing.

Cricket First rules 1774

Cricket, the world's second most popular sport (the British game of football or soccer as the Americans call it, is the world's most popular sport), originated in rural England in the 16th Century and is now played in most countries across the world. Often derided and frowned upon by our cousins across the pond (USA), ironically the world's first international cricket match was played between the USA and Canada in 1844. Now, according to the ICC (International Cricket Council) registrations, more than 107 countries are registered and actively compete and play the world's second most popular sport; cricket.

Darts Traditional pub game. 1896

Brian Gamlin devised the dartboard numbering layout. Some form of the game of darts has been in existence for many hundreds of years in Great Britain, but it was Brian Gamlin a British carpenter from Lancashire, who developed the numbering layout that is still used to this day. He devised the seemingly random layout of numbered segments to punish inaccuracy from players.

Football FA 1863

The Football Association (FA) was formed in 1863, although it flourished in England from the 8th century, but under the more recent association directive from the 'Cambridge rules'. The FA was formed during an historic meeting at the freemasons tavern in London's Great Queen Street, chaired by London solicitor Ebenezer Morley, who was a huge advocate of formalising the rules of football, similar to the way Cricket rules had been formalised. At the meeting with Morley, were 12 London area football club representatives, there for the sole purpose of determining a strict code of rules to help regulate the game.

The only surviving club from that meeting is Civil Service FC, a member of one of the lower southern league amateur divisions. The FA is the world's oldest and original football association. Football is now the world's most popular sport by a huge margin.

Football the FA Cup 1872

The world's oldest football competition was first played in 1872 and known as the FA cup. The first ever game took place in that year between Scotland and England and played in Glasgow.

Formula 1 motor racing 1946

Formula 1 is a racing car sport to which all participants must comply with a strict set of rules, hence the prefix formula. The sport known as F1 was planned in the 1930's, buoyed on by European success in motor sport, but was put on hold due to the onset of WWII. The concept was resurrected in 1946 and the first world championship race took place in May 1950 at Britain's Silverstone racetrack. F1 is now the most popular and richest motor sport in the world and is headquartered in the UK. Three quarters of F1 teams design and build their cars in Britain.

Figure skating 1772

The world's first ice skating association was formed in Britain in 1772. The art of figure skating involves either singles or pairs of ice skaters performing jumps or spins, footwork and lifts, in a style akin to ballet dancing. Briton Robert Jones wrote a paper on the art in 1772 called 'A treatise on skating'. The sport has now developed worldwide and has been an Olympic event since 1908.

Golf 15th century

There is and always will be much debate about the origins of the game of golf. However, the modern-day game can be traced back to its British origins in the 15th Century. It was known as the second game of kings with King James IV documented as playing the game regularly in 1500 and Mary Queen of Scots played a game just prior to the death of her husband Lord Darnley. King James VI allegedly played a game shortly after ascending to the throne at Blackheath in London in 1603. The Honourable Company of Edinburgh Golfers established the first standardised rules of modern golf and the oldest golf club in the world is St Andrews (1754). Golf is now the 10th most popular international sport in the world.

Hockey 19th century

The game known as hockey, was played in English private schools in the early 1800s. In 1886 the Hockey association was formed with seven London clubs joining and at that point, rules were drawn up to form the basis for the modern game that we know today. Hockey spread throughout the British Empire with the advent of its popularity amongst the British Army. The International Hockey Federation (IHF) formed in 1924 and field hockey was introduced for the first time as a men's only game at the 1924 summer Olympics.

Horse racing Thoroughbred 1750

In Britain, horse racing is the second biggest spectator sport in the country, next to football. It is one of Britain's oldest sports and has been established since Roman times. Most of the sports rules, regulations and traditions have been developed in Britain, including the establishment of the Jockey Club in 1750, who created the rules of racing. Britain has been the world centre for thoroughbred racing for many centuries and boasts some of the world's most famous racecourses such as Ascot, Cheltenham, Windsor and Epsom. There are two forms of horseracing in Britain: flat racing, run over distances of up to two miles, 5 furlongs and 159 yards and National Hunt Racing: run over distances of up to four and a half miles with horses jumping over hurdles or fences. The Cheltenham Festival is the world's number one jump racing festival.

Ice hockey Shrewsbury 1818

Contrary to common belief, Charles Darwin observed ice hockey and noted in a letter to his son William whilst reminiscing, that he played hockey on an 'iced over pond' on 'ice skates' at school. He attended Shrewsbury School between 1818 and 1825. This was revealed in a book called 'Origin of Hockey" by two Swedish colleagues Carl Gilden, Patrick Houda and French-Canadian Jean-Patrice Martel in 2014.

Netball 1890s

The British sport of netball emerged from early versions of another British game called basketball; a sport invented by British doctor James Naismith and first played in 1891. The game of netball developed shortly after basketball and was initially played by women. Although not technically intended, the game of netball is still dominated by female players the world over.

Olympic games London

London is the only city to have hosted the summer Olympic games three times: in 1908, 1948 and 2012. Despite Britain covering less than 1% of the world's landmass, it ranked third in the world in the 2012 medals table, just behind the USA and China. But the best was yet to come; at the 2016 Olympics in Rio, Britain ranked an outstanding second overall in the medals table behind the USA.

Paralympic games 1948

The first Paralympics games for disabled athletes were held in London in 1948. German born Dr Ludwig Guttmann from Stoke Mandeville hospital, held a competition for war veterans with spinal cord injuries. The games were originally called the 'Wheelchair Games' and were purposely planned to run the same summer as the 1948 Olympic games. The games were held again in 1952, this time with Dutch veterans taking part alongside the British participants. It became the first international competition of its kind. Then in 1960 and no longer the preserve of war veterans, the world's first official Paralympics games were held and attended by 400 athletes from 23 different countries.

Polo Early 19th Century

The game of polo is a team sport played on horseback, formalised and popularised in Britain in the early nineteenth century. The aim of the game is to score points or goals against the opposing team by hitting a small wooden or plastic ball into the other team's goal, hit with a long-handled mallet. The polo pitch is traditionally 300 yards long by 160 yards wide and each team has four riders and horses. The professional sport of polo is played in 16 countries.

Rounders 15th Century

The game of rounder's is the origin for the game now known as baseball, but rounders has been played in Britain since Tudor times. It is a bat and ball game using a hard leather ball, similar in size to a cricket ball hit by a mildly tapered tubular wooden bat. The team members can score by running around the four bases on the pitch. One full round counts as one point. The game is very popular among British and Irish school children. In the USA, adults play the very similar game of baseball professionally.

Rugby league 1895

The game of rugby league was formed as a separate game from rugby union following a disagreement of paying players. Its history dates to 1895, when a split in rugby union teams resulted in the formation of rugby league paid professional players, and rugby union non-paid amateurs. Slowly, the two games evolved into distinctively separate games. Rugby league in Britain is mainly a northern game consisting of 13 players.

Rugby union 1823

Legend has it, that British Public Schoolboy William Webb Ellis started this international sport while playing football at Rugby School during a match one day. In the heat of the moment, he picked the ball up and ran with it into the goal line. The rest is history. Rugby union is now played with 15 players and is the main sport in both New Zealand and South Africa and played enthusiastically worldwide.

Rugby union was completely amateur and non-salaried until a split and argument over payment of players in 1895 erupted. This dispute led to the split of rugby union and formation of rugby league, where all players were paid professionally and from that point on, the two games have evolved with two different sets of rules. The embargo of paying rugby union players was not lifted until 1995 and only then was professionalism sanctioned by the games governing body. A derivative of this game is American football, although the players of rugby union and league players do not wear the body protection that US American footballers don.

Sailboard and sail boarding 1958

Briton, Peter Chilvers built the world's first sailboard. Chilvers was a 12-year-old boy from Hayling Island in Essex at the time of his invention. As a child, he enjoyed most water sports and one day while at the beach near his home, his inquisitiveness got the better of him and he mounted a sail on his surfboard. Chilvers became heavily involved in developing and getting the sport internationally recognised. The sport grew in popularity so fast around the world that it became an Olympic sport in 1984.

Snooker Origin — Mid 19th Century

Colonel Sir Neville Chamberlain bestowed the name 'Snooker' on the game while serving in the British Army in India. He commented that a young player was 'a real snooker', a common term given to a first-year cadet, referring to their inexperience at the time. Snooker is a derivation of the original table game of billiards, a sport with just two cue balls and an object ball. However, using the same size table, snooker uses 15 red balls, six coloured balls and a white cue ball.

Squash — 1820

The game of squash was first devised and played at British public-school Harrow, in around 1820. Pupils discovered that a deflated tennis ball squashed on contact with a wall and this was the catalyst for a game that developed a larger selection of shots, that needed more effort for the players, rather than waiting for a ball to bounce in simple tennis. Squash is now an international and Olympic sport played by millions of players worldwide.

Table tennis or ping-pong — 1880s

The history of ping-pong or table tennis originated in Britain and probably evolved as an indoor version of the British game 'royal tennis' or real tennis, widely played in the Medieval era. There are also suggestions that the game was once played by British Army officers in India and South Africa and known as indoor tennis, using the lids from cigar boxes and balls moulded from cork wine stoppers. But in 1900, Briton James Gibb brought back hollow celluloid balls to Britain from the USA, in the form that are now used for the sport of table tennis the world over. In 1901 both the Table

Tennis Association and the Ping-Pong Association were formed in Britain.

Tennis 1874

Solicitor, Major Harry Gem from Birmingham, devised the sport of tennis with a Spanish friend and two doctors in 1874. Gem then founded the world's first lawn tennis club: the Leamington Tennis Club. Tennis is now the world's fifth most popular sport in the world. Two or four players can play the game either singly or in pairs, played predominantly on a grass court, or in some warmer climates, on clay-based courts.

Tiddlywinks 1888

The game of Tiddlywinks can be traced back to late Victorian times when inventor Joseph Fincher applied for a concept patent application. Followed a year later by a patent filed by Fincher for the trademark of the game, now called 'Tiddlywinks'.

Stonehenge

Tourism, cities and historic sites

10 Downing Street London 1664

Known colloquially as 'number 10' in Britain, 10 Downing Street is the British Government Headquarters and the official residence of the Prime Minister. Forming part of an impressive Georgian terrace, it was first used in 1684 and built by Sir Christopher Wren. Next door at number 11 Downing Street is traditionally the base and home of the Chancellor of the Exchequer.

Avebury Stone Circles c2500BC

Avebury is a UNESCO World Heritage site, situated on the Marlborough Downs. It is the biggest stone circle in Europe, built in prehistoric times.

Balmoral Castle 1862

Set in the lush green landscape of Aberdeenshire, Balmoral castle is a large estate house in Royal Deeside. Balmoral is one of the

Queens royal private residencies, used in the past by Queen Victoria and Prince Albert. It is a working estate that contributes to the local economy and aims to safeguard the environment in its farming practices. Built in 1862 in its present form.

Bath Abbey 1090

Over the centuries, three different churches have stood on the current site of Bath Abbey. Most recently, Sir George Gilbert Scott, who completely transformed the interior, corresponding with his vision of a Victorian Gothic design, built the current spectacular stone vault that can be witnessed today.

Beaumaris Castle 1295

Said to be the most technically perfect example of a castle in Britain, Beaumaris castle was built with perfect symmetry and traditional proportions. Unfortunately. the castle was never completely finished, due to financial constraints with funds being diverted to the ongoing war effort at the time, but most visitors would be unaware because the castle was simply reduced in size.

The Houses of Parliament. Westminster Abbey 1072

The Palace of Westminster encompasses what most people know as the Houses of Parliament and Big Ben and is one of the world's most iconic buildings. It not only houses the famous House of Commons and House of Lords, but its famous clock tower houses the huge cast bell, known more commonly as Big Ben. The origins of the British Parliament date back to the 11th Century.

Blenheim Palace 1705

This stunning palace is the only non-royal palace in Great Britain. It is home to the eleventh Duke and Duchess of Marlborough and famously the birthplace of Sir Winston Churchill, set in more than 2000 acres of beautifully manicured 18th century park and gardens. Blenheim Palace is a World Heritage Site. Prior to 1705, the original building was known as Woodstock Manor.

Buckingham Palace 1761

Serving as the official London residence of British sovereigns since 1837, Buckingham Palace is the official administrative headquarters of the Monarch. The palace is used for many state and official events throughout the year.

Caernarfon Castle 1283

A beautifully preserved medieval castle located in Caernarfon, Gwynedd, North Wales. Town perimeter walls were also part of the huge construction when the castle was built and as a result, the castle was not completed until 1330.

Dover Castle 10AD

Situated in a spectacular setting, high above the White Cliffs, Dover Castle has defended British shores for more than 2000 years. A potent medieval stronghold, and home to the intelligence centre for the Evacuation of Dunkirk in WWII, this beautifully preserved castle has witnessed the shaping of modern British history.

Edinburgh Castle c1130AD

Situated on top of what is now an extinct volcano, with stunning views across Edinburgh city, the castle has been party to many of Scotland's defining events over the centuries. In its current form, the castle was built by David I, son of Saint Margaret of Scotland, around the time of 1130AD. Edinburgh Castle is now one of the most popular tourist attractions in Britain, boasting more than a million visitors a year.

Hampton Court Palace 1494

In 1494, the Lord Chamberlain leased and modernised the medieval gem known as Hampton Court Manor. It was later extended and modernised by Thomas Wolsey, who converted it into a grand palace, often frequented by King Henry VIII. Then in 1689, Sir Christopher Wren demolished large parts of Hampton Court's Tudor palace to prepare for a new palace for King William III and Queen Mary II.

In 1838, Queen Victoria opened the gardens and state apartments free of charge to the public. Hampton Court Palace is now one of the most visited historic attractions in Great Britain and it also boasts the country's largest maze.

Kensington Palace London 1689

This majestic palace has been a favourite home of royals for many years, including, Diana, Princess of Wales and Princess Margaret. Kensington palace is now a much-transformed fashionable and truly stunning residence, now used primarily by the Duke of Wales, Prince Charles and his wife the Duchess of Cornwall.

Leeds Castle Kent 1200AD

Formerly, the site of a Royal Saxon family manor, Leeds castle was built as a Norman stronghold, becoming a Royal palace and home of kings and queens for more than 300 years. The castle is a quintessentially English moated Norman Castle and one of the finest specimens of its kind in the country.

London 2014

Tourism to London increased by 20% between 2013 and 2014, making London the most visited capital city by tourist's numbers, in the world. With its profusion of free entry, world class Museums, theatres, low crime rate and more green space than any other European city, it becomes more attractive and compelling each year, with visitors travelling from all four corners of the globe.

London the world's busiest financial hub.

The 'city' area of London is made up almost exclusively of financial service companies. London lays claim to be the world's busiest financial centre by transactions per day and for the number of financial institutions operating in the financial district. London is also the only financial centre outside of China to have an agreement to trade with the Chinese currency and stock in RMB or Yuan.

London World's only megacity between 1831-1925

Today, when large cities are discussed, megacities such as Tokyo, Shanghai and New York inevitably spring to mind. But during the period between 1831 and 1925, the world's largest city was London with a population of between 4 to 6 million. Today, with its population fast approaching 10 million; London is still the only mega city in Europe with a static population of 9.8 million and a transient population in excess of 16 million.

London First electric advertising 1908

The world's first electric advertising began in central London's famous Piccadilly Circus. The first advertiser at the landmark site was the French company, Perrier.

Roman Baths c70AD

The city of Bath was developed around its natural hot springs, its steaming water playing a pivotal role during its long and rich history. The Roman Baths in the heart of the city were developed around 70AD and formed a hub where high-ranking Roman soldiers participated in social bathing on a grand scale. The baths remain one of the best-preserved examples in the world.

St Michaels Mount 12th Century

Steeped in myth and legend and known locally as the Jewel in Cornwall's crown, St Michaels Mount stands proud on a spectacular island, very close to the mainland of Penzance. It is linked to the mainland by a tidal causeway and boasts a scenic harbour and picturesque gardens. The main building on the Island started its life as a Benedictine Chapel and dates back to the 14th Century. The 'Mount' is the most famous landmark in Cornwall and is now managed by the National Trust.

St Paul's Cathedral 1710

St Paul's famous dome, still dominates the London Skyline. Designed in 1710 by Sir Christopher Wren, it is a truly magnificent architectural work of art. A cathedral has been on the same site since 604AD, but the current masterpiece is dated 1710

Stonehenge circa 3000BC

The UNESCO world heritage site of Stonehenge is a 5000-year-old monolith consisting of giant rings of stone pillars and cross bearers and pre-dates the pyramids by approximately 1500 years. The monument is built and aligned to the setting and rising of the sun at both solstices, although its exact function remains an enigma.

Tewkesbury Abbey 1087

Situated between the Malvern Hills and the Cotswolds, Tewkesbury Abbey is an architectural gem that can trace its roots back to 715AD when it was the site of a Saxon Abbey. The abbey is Britain's second largest parish church and one of the finest examples of Norman architecture to be seen anywhere in Europe. The Abbey also boasts Europe's largest Romanesque crossing tower.

Tower Bridge 1886

Built in 1886, the iconic Tower Bridge is crossed by more than 40,000 people each day. The bridge is an engineering masterpiece, combining a suspension bridge with a bascule bridge and crosses the River Thames, close to the Tower of London, hence the bridges name. Tower Bridge has the capability to open and close to allow large ships to pass under its iconic structure. The bridge is recognised the world over as a British icon.

Tower of London 1066

This magnificent castle is located in the heart of London and is home to the crown jewels, guarded by the Yeoman of the Guard known as the famous 'Beefeaters'. The white tower was built in the 1080s and in the 1220s; Henry III commenced major expansion of this royal palace. The last hanging at the Tower took place on Tower hill in 1780. The Crown Jewels Exhibition opened in 2012. The Tower of London is the most visited Tourist attraction in London.

Universities Oldest in English-speaking world. 1096

Oxford (1096) and Cambridge (1209).

Oxford University is the world's oldest University in continuous use. Both Oxford and Cambridge lay claim to be the first Universities still in continuous use in the English-speaking world. Both universities are still ranked in the world's top three after all this time. Interestingly, one of Cambridge's scholars was so inspired by his education there, that when he got back to America he established the first college in the USA in Cambridge Massachusetts, now known as top ranking Ivy League establishment, Harvard University.

Warwick Castle 1068AD

Developed from an original structure built in 914AD and extended by William the Conqueror in 1068, Warwick castle is medieval and one of Britain's favourite tourist attractions. The castle has a moat around the perimeter and is situated on a bend of the River Avon in Warwickshire's County Town of Warwick. It is often said that Warwick Castle is Britain's ultimate castle.

Windsor Castle 1070

The world's largest inhabited castle, standing on the bank of the Thames. Windsor Castle is a royal residence in Windsor in the royal county of Berkshire, originally built in the 11th century. Windsor Castle is also the longest occupied palace used by monarchs in Europe and today, more than 500 people work and live in its grounds. The castle is situated in a picturesque location, close to the banks of the River Thames.

Transport

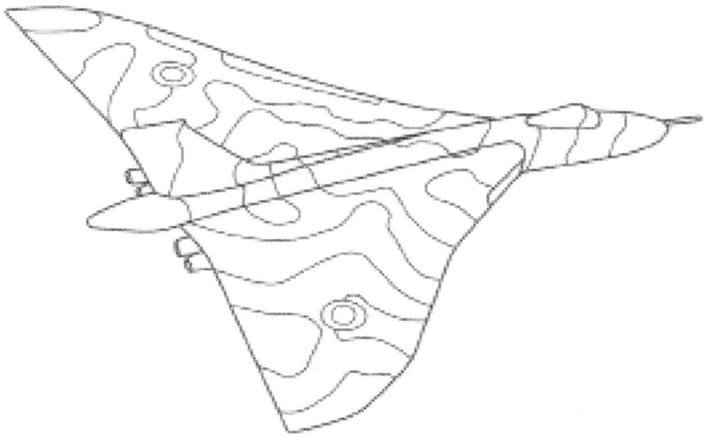

Avro Vulcan

Aviation

Aerial refuelling 1935

Aviation scientist Sir Alan Cobham, was a born in 1894 in London. His illustrious career spanned WWI as a pilot and he once flew to Melbourne Australia from Britain in a De Havilland DH50 Floatplane. In his pioneering work on air-to-air refuelling, Cobham's early research was based on specially adapted Airspeed Courier planes. He successfully developed and mastered the in-air refuelling technique during the latter part of WWII and his design was successfully taken up by the RAF and US air forces. His old company Cobham PLC is still active. This invention made possible the world's longest non-stop mission for a bomber by the RAF's Avro Vulcan, during the Falklands conflict, flying 8000 miles.

Ailerons 1868

The aileron was first developed and patented by scientist, Matthew Watt Boulton in 1868. The patent was based on his academic paper 'On aerial locomotion'. Then 38 years later, the US patent office granted a comprehensive patent to the Wright brothers for their identical version of Bourton's design called 'A system of aerodynamic control'. Substantial litigation ensued over the legal aspects of lateral roll control. The US government eventually passed a legal resolution in the Wright brothers favour in the US, just before the onset of WWI.

Aircraft First Powered flight 1848

John Stringfellow was a British inventor who travelled by air at just under 10 metres. His contraption was then known as the 'Ariel steam carriage'. He was based in Sheffield before moving to Chard in Somerset to pursue a career in the lace industry, but had always dreamed of starting an intercontinental airline. Stringfellow set up a business with an engineer called William Henson to fulfil this dream.

Unfortunately, despite their best efforts at producing glossy publicity brochures and holding many pre-sales forums, they failed to attract any financial backers who were interested in working with them.

Henson parted company at this stage, but Stringfellow persevered and designed optimal wing designs and new lightweight construction methods, whilst simultaneously calculating wing lift to weight ratios for his various wing shapes. Eventually he built a working test model using optimal wing design and powered by a twin propeller steam engine. The frame of the aircraft was extremely delicate due to the need to make a lightweight structure to ensure lift. The vulnerability of the silk covered wings and bamboo frame struts meant that it could not be tested outdoors as the silk became quickly sodden with moisture in an external environment.

Therefore initial tests took place inside a large Chard based milk mill in Somerset. Eventually in 1848, his unmanned, powered flying machine lifted off the mill floor at speed and travelled a mighty

10 yards (just over 9 metres) and it became the world's first powered flight in a fixed wing aircraft. A model of this historic aircraft is now on display at the Science Museum in London.

Aircraft First heavier than air flight 1849

Sir George Cayley (the 'father of aerodynamics') was a British inventor who designed and built a triple winged glider in Brompton Yorkshire, carrying a 10-year-old boy. This feat became the world's first manned flight by a heavier than air craft. Sir George was a prominent MP and extraordinary inventor who also worked on the development of caterpillar tracks, early seat belts and lifeboats.

Aircraft First manned, controlled flight 1853

Sir George Cayley was the first inventor to shift from the idea that all aircraft needed wings like a bird. He designed and built the world's first full sized controllable glider, capable of carrying a fully-grown adult and it was test flown by his coachman over Brompton Dale in Yorkshire in 1853. On landing, records suggest that his coachman immediately tendered his resignation.

Aircraft First Powered controlled flight 1899

Percy Pilcher was a British inventor who set an unpowered gliding record of 250 metres in Eynsford, Kent and came extremely close to becoming world famous by upstaging the Wright brothers record by 13 years! Following his record manned glide, he then focused on powering his 'hawk' glider and decided on developing a triplane three-winged design, because the additional weight of the third wing did not add significantly to the overall weight of the craft, but did make a huge difference to its lift. He used a small petrol powered internal combustion engine to power the aircraft.

On September 30th 1899, he arranged a demonstration of his new powered aircraft at a venue near the town of Rugby. Unfortunately, the craft was not flight ready. The day was stormy and very windy and rather than disappoint the invited crowd, he decided to fly his glider, the Hawk. In the event, the aircraft proved very difficult to control in the gusty conditions and the tail plane snapped off while flying, sending the craft plunging to the ground, killing Percy Pilcher in the process.

Unfortunately after his untimely death, although almost complete and ready to fly, work was halted on his powered aircraft and dismantled for scrap. However, in recognition of Pilcher's achievements, a BBC programme team worked with Cranfield University in 2002, to build a replica of his powered tri-plane to prove it could fly. Once built, it flew for more than a minute and proved a far better aircraft than the Wright brothers 'Flyer'.

Aircraft Landing on a moving ship 1917

Squadron Commander E.H. Dunning of the Royal Naval Air Service made the world's first aircraft landing on a moving ship. He landed his Sopwith Pup on to the deck of HMS Furious in Scapa Flow, Orkney on 2nd August of 1917. Alas, five days later he was killed while attempting a second landing on the ship that day; his wing tip was caught by an updraft, tossing his plane off the port side of the ship, rendering Dunning unconscious and he drowned, unable to break out of the cockpit.

Aircraft World's first turbo prop airliner 1948

British designed and built; in 1948 the Vickers Viscount was the world's first turboprop airliner to enter commercial service. Within a few years, more than 450 were built and sold worldwide to many countries including the USA and the Far East.

Aircraft World's first VTOL 1954

The Rolls Royce thrust measuring rig (TMR), made the world's first vertical take-off and landing (VTOL) in this year. It was nicknamed the flying bedstead. This development led to the first vertical take-off and landing engines used in the British made Short SC.1 in 1957, using four engines for lift and one for forward thrust. However, this aircraft, although a capable VTOL craft, had very limited flight range due to its five engines

Aircraft Vertical take-off and landing 1965

The British designed and built Hawker Siddeley Harrier jump jet is the World's first commercial and most successful VTOL (vertical take-off and landing) aircraft ever made. The Harrier was the first VTOL to use a single engine (Bristol Siddeley Pegasus) with pivoting nozzles that directed thrust for both forward and vertical flight.

The Harrier is still in service worldwide, primarily flown in short take off and vertical landing mode (STOVL) that enables the craft to carry a higher weapon or fuel load. Designed and launched commercially, initially for the British Navy whose fleet was fully retired in 2010. The United States Marine Corps and the Spanish and Italian navies still use the AV-8B Harrier 11.

Aircraft Commercial supersonic airliner 1969

The Anglo-French Concorde broke the sound barrier and accelerated to Mach 2 (twice the speed of sound). Concorde employed the world's most powerful jet engines in a commercial passenger aircraft, taking it to speeds of up to 1,350 mph and it entered full commercial service in Great Britain on January 1976 between London and Bahrain.

The Rolls Royce Olympus engines each used 'reheat' technology, adding fuel to the final stage of the engine, producing the additional power needed for take-off and the acceleration up to and beyond supersonic flight.

Concorde's fastest transatlantic flight was on 7th February 1996, completing the New York to London flight in 2 hours 52 minutes and 59 seconds. Concorde only ever saw service with two airlines – British Airways and Air France. BA withdrew Concorde in 2003 following the fatal Air France Concorde crash in Paris and the remaining BA fleet was retired for posterity and preservation at airfields in Barbados, Edinburgh, Filton (Bristol), Manchester, New York, Seattle and Heathrow.

Aircraft The world's first autonomous aircraft 2009

The Mantis, thinks for itself, decides for itself and flies by itself. The mantis is a twin engined unmanned aircraft designed and built by BAE systems for the UK Ministry of Defence, developed from concept to first flight in just 19 months. Its first flight was on October 21st, 2009 at the Woomera test range in South Australia. It has a wingspan of 22 metres and uses 2 Rolls Royce M250B-17 turbo shaft engines.

Mantis has a maximum speed of 345mph and can cruise at 230 mph. Its maximum endurance is 30 hours per flight. The Mantis is designed to be used for both reconnaissance and attack operations and can be armed with six Hardpoint missiles. It is the world's first plane to be able to pilot itself and plot its own course, whilst communicating with ground staff to share its observations and strategic data.

Airport Busiest single runway airport

The world's busiest single runway airport and Great Britain's second busiest airport overall is London Gatwick, situated 29.5 miles south of central London and 3 miles North of Crawley town centre in West Sussex. Gatwick started life as an aerodrome in 1930 and in 1933 was approved for Commercial flights by the Air Ministry. In 1958, Queen Elizabeth II flew into Gatwick to officially reopen the newly renovated airport complex and achieved a passenger throughput of just 186,000 passengers in that year. At this time, Gatwick was also the world's first airport to have a direct railway and trunk road link and was also the first with an enclosed pier based terminal, allowing passengers to walk under cover to waiting areas or gates, near to the aircraft.

Gatwick is by far the world's busiest single runway airport with only San Diego airport in the USA coming close with about half the number of annual movements on its single runway. Gatwick's busiest day ever on August 22nd, 2008, saw 895 aircraft movements in twenty-four hours using its single runway. Total yearly passenger numbers are now in the region of 35 Million per year spread equally across its two terminals – north and south. Its single runway is 3,316 Metres long in length and 45 metres wide. It is still one of the few airports in the world with an aircraft stand able to accept the world's largest commercial passenger airliner, the Airbus A380.

Air force Royal Air force 1918

Britain's RAF is the world's oldest independent air force, being formed on April 1st 1918. At the time of its formation, it was the largest air force in the world and it is most famous for its role in its successful campaign of the battle of Britain, fighting the huge forces of Nazi Germany under the command of Hitler. The German Luftwaffe hugely outnumbered the RAF, yet the battle was won on courage, skill and superior aircraft.

Ejector seat 1940

Sir James Martin was an aeronautical engineer who founded 'Martins Aircraft Works' in 1929. He forged a close friendship with test pilot, Captain Valentine Baker. On 12th December, Captain Baker was tragically killed during a test flight of the Martin-Baker MB3 prototype. His death affected Martin so much, that his focus immediately turned to pilot safety. In 1944, Britain's Ministry of Aircraft Production invited Martin to investigate the practicality of developing a means of assisted escape for pilots in all their fighter planes. After many designs, he decided that the most practicable and viable scheme was that of a forced ejection of the seat with the pilot still strapped in place. He decided that the best method for this was to use an explosive charge. The modern-day ejector seat was born and the Martin Baker ejector seat design is now used the world over and to date has saved the lives of more than 7000 pilots and aircrew.

Head-up display 1942

Britain's Royal Aircraft Establishment designed the first prototype head-up display that was built by Cintel. The first system was successfully installed into the famous fighter/bomber, the Blackburn

Buccaneer in 1958. The same basic design is still used in modern fighter aircraft and is now even offered as an option in some prestige cars.

International daily air passenger service 1919

The world's first daily air international passenger service, started life on a grass airstrip in Hounslow Heath, just a few miles from present day Heathrow Airport. The service ran daily between London and Paris using converted De Haviland bombers from WW1. The air service was run by Air Transport and Travel Ltd and the airline was named, Imperial Airways, ultimately becoming British Airways in later years. The Hounslow Heath Airport boasted the world's first arrivals and departure halls and the largest control tower in the world at the time.

Jet airliner World's first jet airliner 1949

The De Havilland Comet was a revolutionary new streamlined jet airliner, the like of which the world had never seen before. The Comet had its inaugural flight in 1949, featuring a pressurised fuselage for safe high altitude cruising. Each Comet came with 4 De-Havilland Ghost turbojet engines hidden in the wings. Several versions of this aircraft were made and the most radical version was the Hawker Siddeley Nimrod, the world's first Jet maritime reconnaissance aircraft. Nimrod saw service in the RAF until 2011, more than 60 years after the Comets first flight.

SABRE HOTOL Hypersonic jet/rocket circa 2020

The SABRE (Synergistic Air-Breathing Rocket Engine) is the world's first hypersonic jet/rocket capable of working safely in space, allowing the opportunity of HOTOL (Horizontal Take Off and Landing). SABRE is still at the conceptual stage, but under development and early testing by British company Reaction Engines Ltd. SABRE is designed to propel an HOTOL craft at speeds in excess of Mach 5 at an altitude in excess of 28.5km high above the earth.

Wind tunnel 1871

Aeronautical engineer, Francis Herbert Wenham realised that in order to test all the forces and performance of good wing and aircraft design, there needed to be a method to test them first. In Great Britain, the science of aeronautics was developing fast and 50 years before the Wright brothers well-documented flight, British inventors had already designed and tested powered and controlled flight.

Although the value of a wind tunnel is obvious to us today, it was not created before test flights started in earnest. Wenham realised that to advance the science of flight and most importantly its safety, an alternative was needed to the 'finger in the air' testing of old. He had tried a similar design of a whirling arm as sir George Cayley used, but after several disappointments, he convinced the Aeronautical Society to build the world's first wind tunnel. The design was a long 12-foot tube, 18 inches square, with a steam driven fan, blowing air at one end down the tube, to the model aircraft at the other end. The testing went better than expected and Wenham's wind tunnel design was of huge importance to the pioneers of the fledgling British aircraft industry.

> "I am well convinced that Aerial Navigation will form a most prominent feature in the progress of civilisation"
>
> George Cayley 1849

Mallard – Fastest Steam locomotive

Rail

MAGLEV World's first MAGLEV Train 1984

Birmingham International Airport was the world's first site to use a passenger train in commercial service, solely using the MAGLEV system of propulsion. Maglev is derived from MAGnetic LEVitation, a means of propulsion that thrusts a vehicle by the use of magnets instead of conventional wheels. By using an electromagnetic central mono track, an object or vehicle can be lifted and propelled dependant on the metal type offered to the magnetized rail. Attached underneath the vehicle, some metals lift and propel the mass forwards, while others act in the opposite direction and are employed as brakes or to reverse the vehicle, dependant on the polar direction of the rail.

The MAGLEV principle was invented and developed by British scientist, professor Eric Laithwaite. The main advantage of MAGLEV systems is that because the drag and friction of each vehicle is low, the vehicles can travel much faster in a safe and controlled manner.

Whereas the negative aspects are that maintenance is very high and because the system is not efficient in its energy consumption, most MAGLEV systems have now been decommissioned.

Today, there are only two commercial MAGLEV schemes that are still in operation: Shanghai, China and Aichi, Japan. Both are very controversial because of the huge cost of construction, maintaining and running the systems.

Railway Passenger 1825

The Stockton to Darlington became the world's first railway to carry passengers and freight. George Stephenson developed a keen interest as an engineer whilst working as a coal miner in North East England. His management team recognised his evolving skills and allowed him to experiment and build machines at his colliery.

When Stephenson first started work in the mining industry, coal-laden carts were pulled by horses to and from the mine and on to the nearby shipyards. Stephenson decided to use steam-powered engines to replace horses, leading to a succession of pioneering world firsts. The most significant of which was the world's first passenger railway between Stockton and Darlington.

Railway First inter-city 1830

The Liverpool to Manchester railway is the world's oldest 'inter-city' railway. The line has been constantly in use since its opening on 15th September 1830. Initially the link was designed to shift cargo away from the expensive and slow ship canal that ran between the port of Liverpool and the growing industrial hub of industry, Manchester. But within a year, such was its success, that the enterprising owners of the line decided to open it up to passengers and within months, it was carrying thousands of commuters.

Railway 1830

The Canterbury and Whitstable railway became the first in the world to offer a regular steam passenger service.

Railway season ticket 1834

The Canterbury and Whitstable Railway was the first organisation in the world to offer passenger rail season tickets.

Railway signals Semaphore 1841

Charles Hutton Gregory installed the world's first semaphore signals at New Cross Gate on the London to Croydon railway. This proved so successful that his technique was used worldwide on all railway systems until electric signals took their place completely.

Railway signals Electric 1893

The world's first fully automated electric signalling system was installed on the Liverpool Overhead Railway.

Railway tickets 1837

Thomas Edmondson from the Newcastle & Carlisle railway devised and produced the world's first pre-prepared and serial numbered railway tickets.

Railway tracks 1603

Huntingdon Beaumont opened the world's first above ground railway system from Wollaton mine to the city of Nottingham. Long before the advent of steam, electric or diesel power. Horses and ponies pulled the rolling stock on wooden tracks and sleepers.

Railway tracks Cast Iron 1767

The world's first cast iron railway tracks were made and laid on the Coalbrookdale-Horsehay railway, England. The tracks were cast and made by a local iron and smelting works in Shropshire.

Railway tracks Steel 1857

The world's first steel railway tracks were laid on the Midland Railway at Derby. Matthew Kirtley was the local railway superintendent and was instrumental in replacing the cast iron rails that were prone to fracturing and rapid erosion and pushed forward his idea of making the rails from the new 'Bessemer steel' using the world's first steel double-headed rail system, used globally today.

Steam engine 1801

Engineer Richard Trevithick's invention would form the basis for all steam powered trains for years to come. In 1801 he tested a steam car, now known as 'the Puffing Billy'. It climbed the notorious Camborne hill in Cornwall successfully and Trevithick instantly became the first person to power an engine with a piston using high-pressure steam. This single invention was to revolutionise and transform the world of transport and commerce.

Steam locomotive 1804

Richard Trevithick built the world's first steam railway locomotive. It ran on the Pen-y-Darren ironworks railway in Wales and carried the world's first railway passengers pulled by a steam locomotive.

Tilting train 1975

The APT (Advanced Passenger Train) was the world's first high speed tilting train, developed and built at British Rails Derby engineering works. It was tested successfully for the first time on the main rail network in 1975. It was capable of running at speeds of up to 155mph on the existing Victorian designed network, powered by a highly advanced control system that was well ahead of any other train design of its time. After a few minor problems with the tilt mechanism and a slight budget overrun, the state-owned engineering works was informed by the government that there was neither the political will or funding to take the APT project to the next level. Tilting trains now run again in Britain, but are Italian built and ironically based on the same British design from the 1970's.

Train diesel 1976

The British class 43 locomotive known as the HST 125 high-speed train, is the world's fastest Diesel locomotive. Made by British Rails Crewe works between 1976 and 1983. The HST 125's top cruising speed is 125mph (200kmph). However, the world record for a Diesel locomotive was last set and recorded in the Intercity HST 125 Class 43 - 'City of Wakefield', running between Darlington and York on November 1st 1987, when it smashed the previous HST record by running at 148.4mph (238.8kmph). The HS125 is still in service today in both Great Britain and Australia.

Train steam World's fastest 1938

The super streamlined Mallard was and is still the world's fastest steam locomotive. Designed by Sir Nigel Gresley, who had already produced the world beating Flying Scotsman, the Mallard went on to achieve an as yet unbroken world record speed for a steam locomotive of 126mph (202kmph). This record was achieved by the 'Blue Streak' on July 3rd, 1938. It is unlikely that this record will ever be challenged.

Underground railway 1863

The world's first underground railway, known as 'The Tube', opened as the Metropolitan Railway in London from Bishops Road, Paddington to Farringdon Street using steam engines to haul the gas lit carriages, that required air escapes to the street level to ventilate the tunnels of steam and toxic smoke from the trains. The unique underground railway was opened in 1863 and its first tunnels were excavated just below the street level using the 'cut and cap' method. Later extensions to the underground railway were dug much deeper through the clay and chalk terrain of London.

Due to the early use of steam engines, some grand Edwardian town houses were requisitioned and demolished, then rebuilt with just a solid facade to match the terraced houses either side. But the other side of the facade was completely open to the elements to expose the track below ground. This enabled engines to vent and let off steam and smoke. One such house still exists today, used only as a tunnel vent at 23/24 Leinster Gardens in Paddington, London. It is almost impossible to spot this house today where it still stands intact. It was not until 1890 that London's underground system was electrified and then the need for these dummy vent houses was no longer a requirement.

Range Rover – first Luxury SUV

Road

Aston Martin 1913

World famous Aston Martin is the iconic luxury British sports car brand. Robert Bamford and Lionel Martin founded the company in 1913. Aston Martin became associated in the 1950s and 60s with luxury sporting and grand touring cars and is still used after 4 decades, by the British fictional spy character James Bond 007, following the first exposure of a DB5 in the 1964 film Goldfinger.

Bicycle Chain drive 1873

Bicycle designer and racer Henry John Lawson made the world's first chain drive safety bicycle. W ith huge advances in steel tube production, the new safety bicycle was made with a full steel tube frame, with a familiar design that is still used today. Metal was now strong enough to make small chain sprockets and the bike was now light enough for any person to power by pedalling a large cog via a chain to a smaller cog on the rear wheel. Thus the first chain driven pedal bike was invented and put into mass production.

Bicycle Lithium-ion powered lightweight electric bike 2005

The world's first commercially available electric bike, powered by the now ubiquitous Lithium-ion battery technology, was first introduced by pioneering British based International e-bike designers and producers, Urban Mover. Prior to this technology, all electric bikes and cars relied on much heavier Sealed Lead Acid battery technology that weighed 6 times the weight of a similar specified Li-ion battery. The bike was called the Urban Cruiser and became popular across the world, due to its lightweight propulsion system and compact Li-ion battery pack. Now, lithium-ion battery technology powers the majority of e-bikes and electric cars worldwide.

Bicycle Modern 1870

Inventor and gardener James Starley developed and patented the Ariel lightweight bike. The first of the modern frame bikes put Great Britain and James Starley at the forefront of bicycle technology for more than 80 years, earning him the unofficial title of the 'father of the bicycle industry'. The main principles of the bicycle were to place the rider at the correct distance from the ground, placing the seat correctly in relation to the pedals and designing the handlebars so that the rider could produce the greatest force upon the pedals with the least amount of effort. The Ariel was the true global start of bicycle manufacturing in Britain. It was the first bicycle to utilize the 'tension-wheel' using pre-tensioned spokes. A design we all now take for granted and has not changed in more than 140 years.

Bicycle Safety with Pedals 1839

Foundry worker Kirkpatrick MacMillan from Glasgow devised a treadle powered wooden framed two-wheel bicycle and becomes the first mechanically driven bicycle in the world. MacMillan completed the treadle driven bike in 1839. It included wheels with wooden spokes and iron rims, a steerable front wheel and slightly larger rear wheel connected to the treadle pedals via iron rods.

Bicycle tyre Pneumatic 1887

The world's first pneumatic bicycle tyre was made by John Dunlop in 1887. British vet John Dunlop first used his ground-breaking invention, the pneumatic tyre, on a bicycle. At last comfort and safety could be bought in the same package. This proposition was also getting cheaper with the advent of better mass production methods. Surprisingly, the chain was later eclipsed on these bikes with the development of a shaft drive to rid the bike of the dirty and often dangerous chain. But this proved very unpopular and only reappeared as recently as 2005 with a design very similar to the 19th century invention. The shaft drive is still unpopular due to the complexities of changing or replacing a rear wheel, compared to a simple chain driven bicycle.

Bus service First inter-city 1831

Entrepreneurs Sir Charles Dance & Goldworthy Gurney inaugurated the world's first steam powered inter-city bus service between Cheltenham and Gloucester.

Bus Hybrid 1998

British technology company BAE Systems, designed and built the hybrid drive propulsion system that powers the world's largest fleet of hybrid buses in the world. Consisting of an ultra-low emission Diesel engine, powering a generator via a bank of high performance lithium-ion batteries, that in turn drive electric motors in each wheel, using no mechanical transmission. Each system incorporates regenerative braking that feeds power back into the batteries, when braking is enabled. BAE's hybrid bus system is now powering low emission buses across the UK and USA.

Bus Double Decker 1909

The world's first closed top double-decker buses were introduced by the Widnes Corporation in 1909. This effectively doubled the passenger capacity using a similar footprint as the original single deck buses that they replaced. Double Decker buses are now used globally.

Car Powered 1711

Inventor Christopher Holtum demonstrated a horseless carriage in 1711, 170 years before Karl Benz's horseless carriage. He gave demonstrations under the piazzas at Convent Garden in London and the car travelled at 6 miles an hour. This was the world's first powered car. Little is known about how the car was powered, but from the documented demonstrations in front of the public and press, it was clear no horses were involved.

Car Electric 1884

Thomas Parker, a renowned electrical engineer and Victorian inventor, made and tested the world's first electric car and was photographed sitting in the vehicle in a picture revealed by Parker's great-grandson Graham Parker, an ex BBC weatherman. Thomas Parker was also famous for electrifying the London Underground, creating overhead tramways in Birmingham and Liverpool and developing the smokeless fuel 'Coalite'. Sadly, Parker did not pursue his dream of electric cars, a technology overtaken at the time, by the development of the simpler and more powerful internal combustion engine. He died in December 1915.

Car World's fastest 1997

Thrust SSC is a British engineering speed car project that currently holds the world land speed record of 763mph, set on October 15th 1997 in the Black Rock Desert in the state of Nevada, USA. The Thrust became the first car to officially break the sound barrier. Royal Air Force Pilot, Wing Commander Andy Green, piloted the Thrust SSC Supersonic Car to its successful world record breaking run.

Car World's fastest Diesel 2006

At the Bonneville Salt Flats on August 23rd 2006, the same man that was responsible for driving the world's fastest car, Thrust SSC, was behind the wheel of the world's fastest Diesel powered car called the JCB Diesel Max. The car was designed and built by a team of engineers from the British earthmoving vehicle company JCB. Diesel Max reached a top speed of 350.092 mph and created a sensation in the world of Diesel power.

Car World's fastest future car 2019

Bloodhound SSC is the world's fastest supersonic car in development, with the aim of smashing the 1000mph barrier. This British project is based in Bristol and will be driven by RAF pilot Andy Green, already the holder of the current land speed record of 1997, in Thrust SSC. The Bloodhound SSC Super Sonic Car world record attempt will be made in the Kaksken Pan, South Africa, in a large flat part of the desert that is 12 miles long and 2 miles wide. Bloodhound SSC will be powered by a jet engine up to 300mph, then a rocket will take over to accelerate up to and beyond 1000mph. This takes 55 seconds to achieve, at which point air brakes and a parachute will slow the Bloodhound SSC down to 160mph, after which brakes made from steel rotors will then come into force to bring the car to its final halt. The car is 46 feet long and weighs in at a fraction over 7 tonnes, with most of the weight coming from the two engines that produce a thrust of more than 135,000 BHP.

Car First permanent 4-wheel drive car 1966

The Jensen interceptor was a British designed and engineered luxury sports car made by Jenson Motors, favoured by the Princess Royal; Princess Anne, who owned several during its production years in the mid-1960s to early 1970s. The Jenson interceptor was the world's first all-wheel, permanent four-wheel drive car of a non-all-terrain design and the first car to be fitted with anti-lock brakes as standard. It ceased production in 1971. This was a truly pioneering car of its time.

Car Standard anti-Lock braking system 1966

British sports car, the Jensen interceptor was the world's first production car to be fitted with anti-lock brakes as standard. It used the Dunlop Maxaret mechanical system, previously the preserve of lorries, aircraft and racing cars.

Caravan 1885

Dr Gordon Stables was a medical doctor who became the proud owner of the world's first caravan made by the Bristol Wagon Company. Built solely for holidaying, the caravan was called the Wanderer and is still owned by the Caravan Club in Broadway (GB) after a long life travelling tens of thousands of miles. It is 30 foot long and weighs in at an incredible 2 tonnes. It is fitted out with an original china cabinet, bookcase and musical instruments in its opulent interior. Two horses pulled the Wanderer when it first toured the countryside in 1885 and has now been retired for posterity.

Cats eyes Illuminated road demarcation 1934

Inventor Percy Shaw created the 'cats eye', now used globally as a self-illuminating and self-cleaning road demarcation unit. Percy Shaw was a road contractor from Yorkshire, who claimed the inspiration for his invention came when he drove home one night from his local bar and spotted the reflection of light in a cat's eye on the road. His invention was voted the greatest invention of the 20th Century in the Inventor Magazine.

Disc brakes 1902

Inventor Frederick William Lanchester created the disc brake to overcome the common problems encountered by conventional drum brakes, such as poor performance in the wet and uneven wear and tear leading to unreliability. His Disc brakes used brake pads fitted to a calliper that straddled either side of the rotating disc, fixed to the outside of the wheel hub. When the brakes were applied, both pads simultaneously squeezed each side of the disc to slow down and prevent the disc and wheel from turning.

Unfortunately, like so many great inventions, it was not until the 1950's after his death, that Lanchester's idea became commercialised. Disc brakes are now used as standard on almost all cars, motorbikes, buses, lorries and trains.

Driving on the left

The majority of people are right handed and many years ago, when troops went into battle on a horse, they wanted to keep their strongest arm (right) free to hold a weapon. Therefore, if you position yourself on the left, then the enemy hopefully will be on your right and that's why to this day, Britain drives on the left. Strategically, it was always the best side to ride on if in battle. Napoleon was left-handed and despite the majority of his army being right-handed, he demanded that they ride on the right. Rumours abound that this is the reason he lost the battle of Waterloo, because he insisted that all his troops use their left hand to hold their weapons and fight.

Jaguar® Luxury sports car brand 1922

Jaguar is a famous brand of sporting and luxury cars and is now part of the Jaguar Land Rover group (JLR). Founded as the Swallow sidecar company making motorcycle sidecars, it then went on to develop passenger saloons. The company changed its name to Jaguar just after WWII to distance itself from the unfortunate war connotations with the initials SS. In 2013 Jaguar announced plans for the construction of a £100 million-pound research and development centre in Warwick to create the new generation of Jaguar technology and innovation. Work commenced in 2014. The company now exports its cars to more than 60 countries, with its biggest markets being USA, China, Europe and in particular, Great Britain. The models XF type and F type have won worldwide acclaim for their quality, innovation, design and engineering style. The Jaguar XE type saloon is the world's first all-aluminium chassis and body mass-produced car.

JCB® Earth moving machinery 1945

Founded by Joseph Cyril Bamford in Uttoxeter, Staffordshire, JCB is the world's 3rd largest manufacturer of construction, demolition and agricultural equipment. In 1951, Bamford started painting his products yellow and to this day, they are recognised the world over due to the bright yellow colour scheme devised originally for safety and visibility reasons. JCB has 22 factories across Asia, Europe,

North America and South America and its vehicles are sold in more than 150 countries. The term 'JCB' is often used as a generic description for mechanical excavators and diggers, appearing as an official phrase in the Oxford English Dictionary.

Land Rover Off road utility vehicle 1947

Landrover is a British icon with a rich motoring heritage and now part of the Jaguar Land Rover group. Land Rover is designed and built in Solihull, Birmingham and Halewood in Liverpool. Landrover is the world's second oldest 4-wheel drive vehicle brand (Jeep being the first) and the brand is by far the most successful and most prestigious 4-wheel drive brand in the world. The range consists of the Defender, Freelander, Range Rover Evoque, Discovery, Discovery Sport, Range Rover Velar, Range Rover Sport and Range Rover. Used by British and foreign governments globally, for transporting ministers, Royalty and VIP's in safety.

Lorry 1870

The world's first practical lorry or truck designed for carrying freight was designed and built by Briton John Yule to carry ships boilers. The lorry was steam powered and was in use until the turn of the century.

Lorry Articulated 1898

The world's first articulated lorry or truck was designed and built by Thornycroft Company Ltd. The lorry was powered by steam, driving a chain to power the front wheels.

Mini roundabout 1909

The world's first circular junction (Roundabout) for road vehicles was built in Letchworth Garden City. During WW11, Frank Blackmore was a British traffic Engineer who developed the offside priority rule for roundabouts, through his work with the UK Road Research Laboratory. Blackmore also invented the Mini Roundabout, in use now in many countries throughout the world.

Motorcycle First petrol powered 1884

Edward Butler was an inventor who developed and patented the world's first motor powered tricycle. Butler called this innovative vehicle 'the Butler Petrol Cycle. Two Years before Karl Benz demonstrated his car in Germany, Butler revealed his three-wheeled petrol vehicle at the Stanley Cycle Show in London during 1884. Many consider quite rightly, that this was the world's first petrol powered car and also the world's first petrol powered three-wheeled vehicle.

One-way streets 1617

The world's first official one-way streets were introduced and restricted to one-way traffic in the city of London. The trial of 17 streets was an effort by the civic authorities to avoid the growing congestion as London experienced huge population and traffic growth. The experiment was so successful that not only did all 17 trial streets remain one-way, but many more streets were also designated one-way in the following years of London's rapid expansion.

Racing car circuit Brooklands 1907

The world's first purpose built racing car track was built by British land owner, Hugh Locke-King. Work commenced construction in 1906 and the design remit was awarded to Colonel H.C.L Holden who came from the royal artillery. From Locke-Kings original plans for a road track, he was persuaded to revise the design so that participating cars could achieve the maximum achievable speed with the greatest safety levels built in.

The oval circuit was a total length of two and three-quarter miles and one hundred feet wide with two huge banked sections at each end of the oval, to ensure cars could maintain the greatest possible speed in relative safety when cornering. The concrete track was built in a record time of nine months and was one of the most remarkable achievements in civil construction of its time.

Range Rover First luxury off road 4X4 car 1970

Range Rover was designed and crafted by the British based Rover Company, the maker of the world-famous Land Rover. Back in 1966, the Rover design team realised that there was a huge market opportunity for a new segment of a 'go-anywhere' vehicle in Europe, based on US success with the popular but cumbersome Ford Bronco. It was at this point that Rovers chief designer, Charles King, began work on a prototype. Together with his brilliant team of designers, King not only came up with the basic shape of the car, but the team also created the mechanical package needed to power the vehicle.

The chairman of what was then called British Leyland, was so impressed with the concept that he sanctioned the project immediately. The first Range Rover rolled off the production line in 1970 and immediately won acclaim by the international press. Up to 1981, little had changed until an optional four-door model was launched. The first Diesel engine was not an option in a Range Rover until 1986. It was not until 1994 that a MKII was launched, true testament to the original design and longevity.

Range Rover evolved from being a fairly utilitarian vehicle to gradually become more and more luxurious. The MKII was replaced just 6 years later, when it was replaced with an even larger, more luxurious version that stood the test of design and time until 2014 when the MKIV was launched, a mere 42 years from the original MKI model.

Rolls Royce 1906

Regarded as the World's number one luxury saloon car, British Company Rolls Royce is the epitome of design, engineering and quality and is regarded as being peerless in both the world of cars and aeronautical jet engine production. Rolls Royce was founded in 1906 in collaboration between Henry Royce and Charles Stewart Rolls, the result of a partnership in 1904. Their first luxury car was the Silver Ghost in 1906 and became known as 'the best car in the world' due to its new standards of refinement and build quality.

Charles Rolls, an accomplished pilot, was the first person to make a double non-stop crossing by aircraft over the English Channel. Rolls was later killed that year, receiving the unfortunate record of being the first Briton to be killed in an air crash.

Over the years, Roll-Royce launched more new cars and created a luxury sector more opulent than had ever been witnessed before. Rolls Royce had been designing and building aero engines such as the famous RR Griffin and Merlin engines and started designing and producing jet engines after WWII based on a Frank Whittle design. In 1931, Rolls Royce bought the ailing British car company Bentley and after being crippled by development of the RB211 jet engine, Rolls Royce was nationalised. In the early seventies the company was split into Rolls Royce Motors and Rolls Royce Aviation (Jet engines). Rolls Royce Motors was privatised in 1987 as Rolls Royce PLC, bought by Vickers but then sold in 1990 to the Volkswagen Group. In 2003, BMW gained the right to produce cars under the Rolls-Royce name and the new Phantom was released later that year.

Scooter Childs 1912

The world's first commercial child's scooter or Ska-cycle (derived from skate-cycle) was patented by C.E. Richardson & Co Ltd, a company famous in its time for making Cyclecars in Sheffield. Before the First World War, they specialised in children's toys, including toy cars and model aircraft. The company later went on to design and produce more than 600 lightweight low power cars and exhibited each model at the Motor Car Show at Olympia in 1920.

Spare wheel 1904

Brothers Thomas and Walter Davies invented the world's first spare wheel after they saw a gap in the market following several reports of motorists being stranded after accidents or getting punctures. The 'Stepney Spare Wheel', named after the street where their ironmongery and bike shop was based, was an instant success when it was launched in 1904. The Davies brothers sold more than two thousand wheels a month across Europe and the USA at the peak of their business, before car companies decided to make them a standard fitment on all new cars.

Street lighting 1417

It is widely accepted that the Mayor of London, Sir Henry Barton, first ordered illumination of streets in public places by written public order. However, there is no further documentation to detail how this was introduced.

Street lighting Gas 1794

William Murdoch was an engineer and inventor who is acknowledged as the first person to design and install gas street lighting. Murdoch's idea came from his days working in Cornwall on a pump engine installation. Whilst relaxing by an open fire one night, he noticed that after placing a sprinkling of coal dust in his pipe, when placed in the fire, coal gas was produced, coming through the pipe stem and out of the mouthpiece. This was the point that he discovered gas from coal could be used as a lighting medium. He then set about installing a gas lighting system in his house in Redruth and this is recorded as being the world's first house to be lit by gas.

Street lighting 1807

Samuel Clegg was a British mining engineer and colleague of gas lighting pioneer, William Murdoch. Witnessing the potential of Murdoch's original gas lighting system, Clegg travelled back to London and set up his own business called the 'Gas Lighting & Coke Company'. On January 28th 1807, his installed gas lighting system enabled Pall Mall in London, to become the first Street in the world lit at night by street lighting.

Taxi 1625

The world's first documented for-hire wheel-based taxi service related to London Hackney Carriages in 1625. The 'Hackney cabs' as they were known, consisted of an enclosed 2 or 4-seat coach with room at the back for luggage and privacy curtains for its passengers inside the upholstered compartment. The cabs were horse drawn and were generally available to hire from London inn Keepers. The world's first taxi rank was created in London on the Strand in 1635 and in the same year an act of Parliament was passed to make horse drawn carriages for hire legal in the City of London. The first taxi licenses were issued in 1662.

Traffic lights 1868

Inventor John Peake-Knight installed traffic lights at the junction of Great George Street and Bridge Street near the Houses of parliament, London. The lights were operated by gas. Globally, traffic lights have changed the world

Vehicle road tax 1921

Britain introduced the world's first national vehicle tax (VET vehicle excise tax) that applied to all 4-wheeled vehicles. This disc was to prove payment of the tax to use a vehicle on a public road. Ironically, on October 1st 2014, road tax discs were abolished in Great Britain due to advances in electronic registration and camera enforcement.

> "Have you ever noticed that anybody driving slower than you is an idiot, and anyone going faster than you is a maniac?"
>
> George Carlin - comedian and actor

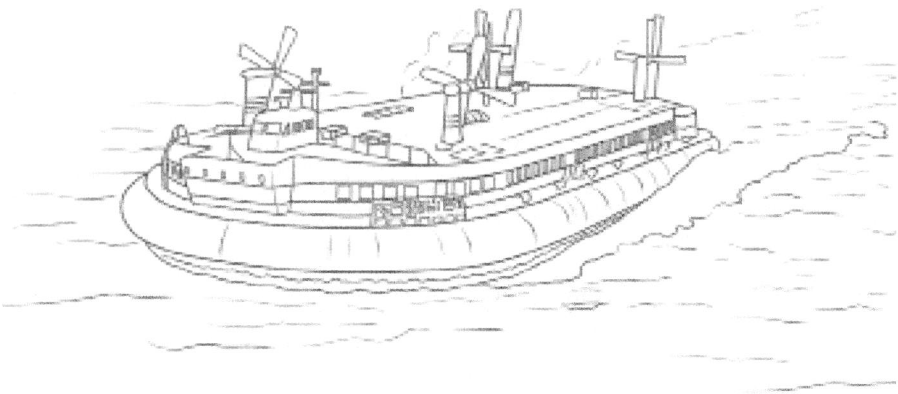

Cross channel Hovercraft

Sea

Chronometer Marine 1761

John Harrison invented the chronometer for accurate navigation at sea. Precise navigation at sea has always been extremely important, but until the development of the marine chronometer, it was almost impossible. It wasn't until 1714 that the British Government offered a huge cash incentive of £20,000 for somebody to solve this problem. British inventor John Harrison had spent his whole working life trying to solve this problem and eventually got his reward for developing the marine chronometer, worth more than £3 Million in today's money.

Diving bell 1691

Edmund Halley was a medical doctor, astronomer, mathematician, physicist and meteorologist. He finalised designs for a diving bell that enabled the user to remain submerged for long periods of time. Halley created a window on the front face so that the diver could observe and explore freely. To replenish air, Halley's design included weighted barrels of air that were sent down at set intervals, allowing the diver to extend their underwater explorations. Halley is best known for calculating a comets orbit that was ultimately named after him, following verified proof and acceptance in the world of astronomy.

Dover Busiest Passenger seaport

The world's busiest passenger seaport Dover is situated on the south east coast of Great Britain looking out across to the coast of France and on to the world's busiest shipping lane, the English Channel. The Port of Dover sees more than 16 million passengers, 2.1 million lorries, 2.8 million cars and motorcycles and 86,000 coaches passing through its terminals each year.

Greenwich meridian 1884

The Greenwich Meridian is the line that divides the eastern Hemisphere with the west and is the official centre of world time. GMT is used to set every single timepiece in the world. GMT was officially adopted globally in 1884.

Hovercraft Amphibious 1953

Christopher Cockerell was a inventor from Cambridge who designed and built the world's first hovercraft called the SRN1. Cockerell owned a small boat building company and was looking at innovative ways to increase the speed of boats by reducing drag. He knew that Thorneycroft had experimented with a prototype 'air boat' where a huge top mounted engine powered a large fan, creating a huge uncontrolled downdraft, lifting the vessel slightly, but was hugely inefficient. Cockerell set about designing a more effective method of uplift by introducing rubber side skirts, containing the air and providing greater lift from a smaller motor and fan, to which he patented his design in 1955. It was built by Saunders-Roe and launched on 30th May 1959.

Hydrofoil 1899

British naval architect and inventor John Isaac Thorneycroft, founded the now famous Thorneycroft shipbuilding company. Thorneycroft developed and introduced many innovations over the following years, but the most significant was his Hydrofoil principal. While working on a method of hull lubrication by air, he experimented with underwater 'wings' to lift the boat and therefore reduce drag, enabling the boat to travel at higher speeds with less power required. This work developed into the first version of hydrofoils called 'stepped chine hulls' that Thorneycroft fitted and used successfully for naval motorboats in WWII.

Plimsol line 1870's

Politician Samuel Plimsoll from Bristol, conceived and developed the Plimsoll line for ships. His invention consisted of a horizontal line on the hull of a ship to denote the maximum safe payload and therefore the safe free space available, depending on operational requirements and conditions. This simple mechanical method of ensuring payload safety in shipping is still used on ships globally.

Ship Steam 1838

The SS Great Western started the world's first transatlantic paddle steamer service. This great craft was designed and built by British engineer, Isambard Kingdom Brunel.

Ship First propeller driven 1838

The Archimedes was a British designed and manufactured ship and when launched in 1838 became the world's first screw propeller driven ship.

Solar powered boat 1975

The world's first documented solar powered boat is a vessel built by inventor Alan T Freeman and named 'solar craft'. Making its maiden voyage on 19th February 1975 on the Oxford Canal at Rugby. The

'Solar Craft' was fitted with ten Lucas solar modules and enabled the 2.5M long catamaran to travel at a lowly speed of 2.5mph.

Steel hulled ship with screw propeller 1843

British engineer and prolific inventor Sir Isambard Kingdom Brunel, designed and built the first passenger liner in the world with a steel hull and screw propeller. The SS Great Britain was at the time the longest passenger ship in the world and built by Brunel for the Great Western Steam Ship Company for their service between Bristol and New York. She was the first iron-hulled liner to cross the Atlantic in 1845 and took 14 days for the crossing. After being rescued from the Falkland Isles where it was being used as a warehouse, the SS Great Britain is now in dry dock in Bristol fully restored and is an award-winning visitor attraction.

> "You can never cross the ocean unless you have the courage to lose sight of the shore."
>
> Anon

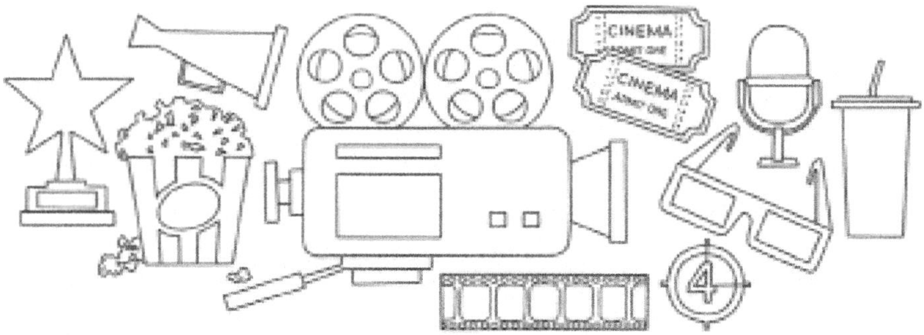

Television, leisure and entertainment

BBC the largest and oldest broadcaster 1922

The BBC evolved from a venture originally set up by the British Post Office in 1922. At the time, it not only controlled mail services, but was also responsible for telephone and wireless broadcasting. Up to this point, government bureaucracy had prevented full scale broadcasting via radio transmitters in Britain, for fear of interference with essential services such as the armed services and police.

By 1922, public attitude and the impressive growth of radio in the USA, forced huge public pressure on the authorities to restore full and proper service. Following much deliberation, the British Broadcast Company set up eight radio stations countrywide, hence the birth of the BBC. In 1932, the BBC had been working with television pioneer John Logie Baird and gave the go ahead for regular television transmissions on an experimental basis. In parallel with the BBC, Marconi Company and the Electronic & Music Industries Co. (EMI) were developing a high definition broadcast system.

In 1934, the BBC adopted the high definition system using 405-lines over Baird's low-resolution alternative 240-line system. The first non-experimental transmission was made in 1936, across the whole of London from Alexandra palace, due to its high position and unobstructed view across the whole of the capital. Leslie Mitchell uttered the first words on BBC television during Britain's public HD television programme debut in 1936.

BBC Journalists

The BBC has the second largest number of journalists (next to China) in the world. The BBC news department is the world's largest news broadcast organisation. The BBC runs 45 foreign news bureaus worldwide, with news correspondents in most countries. In 2014, the BBC celebrated 60 years of television news broadcasts.

Blackpool illuminations 1879

The World's first commercial street illuminations are still operating in Blackpool. Already a thriving seaside resort in the 19th Century, the promenade was lit by Gaslights. But by 1879, the local municipal authority committed £5000 to research and experiment with the new concept of electric street lighting. In September of 1879, an electrical engineer called Doctor Siemens, installed 8 electric lighting devices powered by dynamos and driven by 16 engines, that in turn powered eight arc lamps spaced 320 yards apart and in total produced illumination equivalent to 48,000 candles. These were quite possibly the world's first electric streetlights. The event was advertised nationally and attracted people from not just Great Britain, but other far flung countries too.

In 1912 a display of advanced lighting was commissioned for a royal visit to Blackpool and it was in this year that the chief electrical engineer at the time was asked to provide a string of garland lights to decorate the promenade, using about 10,000 lamps. The event was such a success, that from that year on at the start and end of the holiday season, the now world-famous Blackpool lights were installed permanently to effectively extend the holiday season. To this day, people still travel far and wide to witness the magnificent spectacle of the greatest permanent street light displays in the world. Today, the vast majority of lights used in the displays are ultra-low energy LED (Light Emitting Diodes).

Britain's got talent 2007

The world's most successful TV reality franchise format created by Simon Cowell, is British and now licensed in more than 60 countries worldwide including China and the USA from the 'Got Talent' franchise. It is distributed by Freemantle Media and is produced in association with Syco TV (Simon Cowell TV).

Circus World's first 1780

By expanding a horse demonstration, Philip Astley's shows held in his equestrian centre, grew into the world's first circus. Astley began to incorporate acts from fairs and other shows in London including, ropedancers, jugglers, strong men, jesters and acrobats. In 1780 he constructed a roof over his arena near Westminster Bridge in London to ensure packed houses in all weathers. He is also acknowledged for discovering that the ideal size for a circus ring was a diameter of 42 feet

Coronation Street 1960

Coronation Street became the world's longest running television soap opera on September 17th, 2010, according to the Guinness World Records. The programme started in 1960 on December 9th at 7pm. Coronation Street is centred on families living in a cobbled terraced street in the North West of England, built at the turn of the 20th Century.

Crosswords 1913

British inventor Arthur Wynne was born in Liverpool in 1871. He emigrated to Texas in the 1890s and after a spell of farming became editor of 'the fun section' of 'New York World'. In December 1913 Wynne devised the world's first crossword puzzle (then called 'word cross') and it became an overnight success from the papers readers. Unfortunately for him, he never patented or copyrighted the principle of the crossword puzzle and did not profit from his invention, but he certainly made millions of people happy and frustrated in equal measures across the world to this day.

Game show for TV 1938

The world's first Game show for TV was Spelling Bee televised and broadcast by the BBC from the Alexandra palace in London in 1938. This ground-breaking but simple game show was a competition where contestants were invited to spell a selection of words of differing difficulty levels throughout the show.

Holiday camp 1906

Dodd's socialist holiday camp was world's first seaside holiday Park. Based in Caister near Great Yarmouth, it opened its gates in 1906 and guests stayed in tents, woken early in the morning by large loudspeakers and the camp manager, advising customers of the day's events. This holiday park was the predecessor to Butlin's, Pontins and Center Parks.

Jigsaw 1767

John Spilsbury was an engraver who created his first puzzle in 1767 as a teaching aid for schools, particularly in the subject of geography. He stuck a world map to wood and engraved each country to create his first world map puzzle. He went on to commercialise these puzzles and after his death, his wife Sarah continued the business.

Mousetrap World's longest running play 1952

British born Agatha Christie is the world's most successful author and most successful crime writer of all time. Christie is the playwright of The Mousetrap, the world's longest continually running play with over 25,000 performances in London alone. The play is still running in London's West End.

Package holiday land travel 1855

Businessman Thomas Cook started the 'International Travel Company' in 1841 by launching the world's first packaged day trip by train, from Leicester to Loughborough. Then after years of very successful British organised tours, he developed a packaged travel experience into Europe. He issued vouchers in advance for accommodation, meals and travel from Harwich to Antwerp and onward to several European destinations. This historic event was the world's first overseas package holiday. Now Thomas Cook is one of the world's leading travel companies.

Package holiday air travel 1919

British Travel agent and entrepreneur Thomas Cook became the first company in the world to offer tourists pleasure trips by air. Making this creation, the world's first air package holiday.

Safari park 1966

Jimmy Chipperfield set up the world's first drive through safari park outside Africa, at Longleat manor in Wiltshire. The owner of Longleat, the Marquess of Bath, agreed to fence off about 100 acres of land to accommodate 50 lions and Longleat opened its doors to the general public in 1966. The park is home to more than 500 animals in total and Longleat Safari Park now covers an area of more than 14 square miles (36.42 km2).

Scouting 1907

Robert Baden-Powell was a British army officer who decided to hold a camp for boys on the Brownsea Island in Dorset after which, the following year he published a book called scouting for boys; an adaptation of a military guide that he had previously written. From this point on, the scout movement had begun. The organisation grew rapidly and included girl guides, cub scouts and brownies, and built on the key activities of hiking, woodcraft and camping. Powell established the distinctive uniform with coloured neckerchief and 'woggle', originally introduced to mask the differences in social standing and wealth between the individual scouts. He also introduced badges of merit to recognise achievement in a wide variety of activities. Now the scout movement has more than 40 million guides and scouts in more than 200 countries worldwide.

Teletubbies 1997 to 2001

Teletubbies is a British children's television show aimed at babies and toddlers. Produced from 1997 to 2001, rights were granted for the show to be televised in more than 60 countries from Britain to Brazil and Australia to America. It is even popular in China and South Korea. Devised by Anne Wood and Andrew Devonport, it was filmed near Shakespeare's birthplace, Stratford-upon-Avon. The show featured five brightly coloured characters living on a diet of 'tubby custard'.

Who wants to be a millionaire? 1998

David Briggs, Mike Whitehill and Steven Knight created the original British format. Sony Pictures Television now owns it and the basis of the game is that large cash prizes are offered to contestants whose cash pot grows, the harder the questions get. The maximum prize in Great Britain is £1 Million pounds. In most countries the maximum prize is a million of its local currency. The original version of Millionaire aired on British TV 4th September 1998. It is now the world's most successful TV game show format and is an international franchise of British origin, licensed in more than 100 countries worldwide.

> "Britain has an impressive history of developing formats that click with viewers. Two of America's highest profile and longest running reality shows, " American idol" and "Dancing with the stars," were based on U.K. programs.
>
> John McVay

Nessie from Loch Ness

Tradition and legend

Bonfire night November 5th 1605

Guy Fawkes night or Bonfire night originates back to 1605 when a catholic restorationist terrorist known as Robert Catesby, put in charge Guy Fawkes with his associates, to jointly blow up the Houses of Parliament while in session. The plan was to demolish the Houses of Parliament while the protestant King James I and the entire protestant and surprisingly, the catholic nobility and aristocracy were inside. In the event, government intelligence led to the arrest of Guy Fawkes and his co-conspirators just a few hours before the planned attack. It followed a search of the basement of the Parliament buildings during the early hours of November 5th, provoked by the delivery of an anonymous letter warning of the imminent terrorist attack. Each year thereafter, British people traditionally celebrated the failed attempt to destroy the blowing up the houses of parliament in Westminster by holding celebrations on November 5th each year, traditionally with a large communal bonfire and firework displays.

Cheese rolling 15th Century

Cheese rolling is a traditional event held annually during the spring bank holiday at Coopers Hill between Cheltenham and Gloucester in the English region of the Cotswold's. It was traditionally attended and participated by the local people of the Village of Brockworth, where participants chased a whole round of cheese from the top, to the foot of the hill. The fact is that nobody really chases the cheese because the slope is so severe that everybody tumbles down.

Such is the popularity of this event that competitors now come from all four corners of the world to take part in this often dangerous and exhilarating tradition. The sharp incline of the hill and the extremely uneven surface, often results in a number of injuries, including concussion, broken bones and sprained ankles. Paramedics are always in attendance at the foot of the hill ready for the inevitable emergencies that follow.

Halloween 10AD

Celts celebrate New Year on November 1st. October 31st marked the end of harvest and the summer and the countdown to the start of the long winter nights. October 31st was therefore known as 'the Hallowed Eve', also known as all Hallows day. Contrary to common belief, the Americans did not start celebrating Halloween until the middle of the 19th century when British immigrants took their customs over there with them.

King Arthur circa 500 AD

King Arthur is a legendary British leader and king of the late 5th and early 6th centuries, who, according to medieval legend, led the defence of Great Britain against Saxon invaders in the early 6th century. His knights of the round table famously protected him and there are remains of his cliff top mythical castle in the village of Tintagel on the North Cornish Coast.

Loch Ness and Nessie circa 1000 AD

Loch Ness is one of Scotland's great lakes, located in the north west of the country. The loch or lake is twenty-four miles long and has a maximum depth of seven hundred and eighty-eight feet. The depth and natural darkness of Loch Ness provides a feeling of mystery. People are said to fear what they don't know. This is true to the tradition of Loch Ness, as legend has it that there could be a monster lurking in its depths. Nessie, the Loch Ness Monster is among the most well-known urban legends of Great Britain and it is among the greatest points of interest for tourists to Scotland, although its existence has never been proved for sure.

Maypole dancing May Day celebration

The tradition of maypole dancing was popular throughout Northern Europe especially Great Britain and many Germanic countries since the Iron Age and is still practiced on May 1st in many parts of Great Britain. Dancers run circles round a tall pole holding on to ribbons that are fixed to the top of the mast and the ribbons become intertwined and plaited around the pole. It is then tradition that the dancers perform their set piece dance to unravel the ribbons to their original form. The dancers perform to traditional British folk music played by musicians dressed up in the same bright traditional costumes as the dancers. May Day is now normally the first Monday of May, but traditionally it was always the first day of May.

Morris dancing Middle ages

Morris dancing is a traditional form of English folk dancing normally accompanied by live folk music played by a small band of musicians using traditional instruments. The dancers usually follow choreographed routines using a combination of coloured sticks, swords and handkerchiefs and each dancer normally attaches bells on their wrists and ankles to add to the rhythmic nature of their dance routines.

New Year's Eve 31st December

New Year's Day is a 'Bank Holiday' or public holiday in Britain and is the first day of the Gregorian calendar; a time to reflect the past year and look forward to the new. The celebrations welcome the New Year on the eve, which is the 31st December. In Scotland people celebrate the same festival known as Hogmanay. Throughout Britain there are parties, singing, dancing and fireworks, to 'ring out the old year and ring in the new, traditionally referring to the local church bells that struck at midnight. It is traditional as the clock strikes to sing a song known as 'Auld Lang Syne, referring to friends old and new.

New year first footing — January 1st

In days gone by there was a tradition of 'first footing'. This tradition was meant to bring good luck to all who participated for the coming year. As soon as the bell struck twelve midnight, people queued up by their front doors waiting for a dark-haired person to arrive. This person brought some bread to ensure everybody would have food to eat for the year, money for continued wealth, a green plant to ensure a long life and a piece of coal to ensure that the home would always be warm. The dark-haired visitor would then take an ash pan of dust or ashes and empty it outside the house to mark the end of the old year.

Pearly kings and queens — 19th Century

Pearlie's as they are affectionately known, are an organised group with charitable status based on working class culture in London. Legend has it that pearly kings and queens were first associated with an orphan street sweeper called Henry Croft, who spent his adult life collecting charitable donations for the less well-off citizens of the capital. An official Pearly society was formed in 1911 and members can often be seen on the streets of London dressed in their beautifully hand-made traditional suits, dresses and hats, adorned with thousands of pearl buttons and sequins.

Robin Hood — Circa 1377

The legend of Robin Hood is based on an outlaw who lives his life by robbing the rich and giving to the poor in the area of Nottingham, in the East Midlands of England. He achieves his crimes with the support of his band of merry men and his partner, Maid Marion.

His name is first mentioned in William Langmans play Piers Plowman. Legend has glorified both Robin Hood and Sherwood Forest. There is no firm evidence of his existence, although a criminal by his deeds for centuries, Robin Hood has been the hero of the lower classes.

Saint Patrick c.461–493 AD

Born into a wealthy British family, Patrick's father was a deacon of the Christian faith. Kidnapped by Irish raiders, he was sold into slavery and worked as a shepherd finding solace in his newfound religious faith. He later escaped Ireland, stowing away on a boat back to Britain, later becoming a priest. He then nurtured a dream to return to Ireland as a missionary and despite fraught times with Irish pagan chiefs, he converted many pagans and organised Irish Christians.

Legend has it that he used the three-leaf shamrock to symbolise the holy trinity. He became the second bishop of Ireland and by the 8th century, Patrick was the patron Saint of Ireland. St Patrick's day is celebrated worldwide on 17th March, which according to legend is the day of his death.

Silver ball hurling Cornwall c.1000 AD

Shrove Tuesday is a traditional time in the ancient Cornish town of St Columb Major, when the 'local mob' aim to hurl a small silver ball into one of two goals set roughly two miles apart at either end of the town's main thoroughfare. The winner gets to share several jugs of beer along with fellow competitors after being carried back to the towns market square. Shrove Tuesday is celebrated the day before Ash Wednesday and is therefore the final day before the commencement of Lent, a Christian festival leading up to Easter Sunday (Easter Day).

Conclusion

The metaphor of choice for 'progress' is usually a locomotive: an engine surging forward on a linear course towards a scintillating future. But the path of innovation and invention is almost never linear. Discovery happens on a landscape of competition, fits, starts, explosions, thrusts, failures and grand successes. For much of modern history, the central landscape of discovery was a collection of small Islands in the North Atlantic. For a thousand years, Britain has served as the innovation incubator for the world.

For so many of the technologies that we take for granted as conveniences of the modern world, their story begins in Britain.

Long before the Wright brothers were even born, the science of aviation had been elucidated by pioneers such as Sir George Cayley and Percy Pilcher. The world's first commercial Jet airliner was the Comet, a British designed and built aircraft that launched nine years before Boeing produced a comparative form of transport.

Before Sir Alistair Pilkington invented float glass, the world was a far hazier place. Now, float glass, which allows for uniform thickness and flat surfaces, is the most widely used form of glass in the world.

Britain produced the world's first steam engines, passenger railways and the underground train network.

In medicine, British scientists have helped save many millions of lives with the development of antibiotics, vaccinations, and anaesthetics. Equally important are the many British discoveries in basic biology that have helped lay the foundation for the medications of the future such as monoclonal antibodies that were first humanized by Greg Winter in 1986.

The spread of the English language, for better or worse, has meant that it is now the de-facto lingua franca in science, business, aviation and the Internet.

Britain ranks number one in Nobel Prize winners (132) per capita.

The scale of these discoveries raises one major question: how? How is it that such a small island nation has contributed so much to modernity?

The answer lies in the creation of a system of scientific discovery that is deeply rooted in education and a culture of curious engagement with the world, premised on the idea that human reason can uncover the secrets of the universe.

This system has produced a continuous supply of global innovators and leaders. It should not come as a surprise that Great Britain remains home to two of the top three universities globally (both the oldest in the English-speaking world).

Leading British thinkers have done more than just invent. They created the system of rationality that allows for invention. Roger Bacon, an Oxonian, credited as developing the first formulation scientific method, may be the preeminent example. Bacon (1214–94) was promoting a theory of scientific methodology 700 years before the world's scientists finally accepted it as the dominant form of conducting research.

Throughout the renaissance, Britain continued the tradition of nurturing brilliant minds at the forefront of science and its advancement.

Many British people led at the vanguard of the Scientific Revolution. A prime example is William Gilbert (1544–1603), who laid the foundations for the theory of magnetism and electricity.

Around the same time, William Harvey demonstrated that blood circulates through the human body.

Sir Isaac Newton advanced the law of universal gravitation and through his works, showed that scientific theory should be coupled with rigorous experimentation, which became the basis of modern science.

Britain stood at centre-stage in the Industrial Revolution

that spanned the 17th and 19th centuries. British people and firms helped lead this world historical transformation and fundamentally changed how commercial activity takes place, how products are manufactured and how society is organised.

In the midst of these social changes, British scientists made earth-shattering discoveries of their own: Charles Darwin introduced his radical theory of evolution by natural selection in 1859, and Michael Faraday equipped later generations of inventors and physicists with the principles of electricity, magnetism, motors and generators.

The scientific prowess of British inventors, philosophers and innovators continued right through the 20th century.

Alexander Fleming, the pharmacologist, discovered and developed antibiotics with his discovery of penicillin. British innovators could be found across virtually every field of inquiry: Ernest Rutherford, the father of nuclear physics, John Logie Baird, the father of television, and Alan Turing, the father of theoretical computer science and artificial intelligence.

There has been no deceleration in the rate of British innovation, even in the fully globalised 21st century. The British are still world leaders in modern medicine, engineering, science, music and literature.

In fact, after more than a thousand years of invention and innovation British science is still the envy of the world. When the 2016 Nobel Prize for Physics was announced, it felt completely natural to learn that all three of the recipients were British.

Given the full weight of British achievement in science and technology, it would be reasonable to imagine that humanistic and artistic pursuits would be less emphasised. But nothing could be further from the truth.

British poets and writers have left indelible marks on the world throughout the centuries: William Shakespeare, Virginia Wolf, Byron, George Orwell, Shelley, Charles Dickens, George Eliot, and

JK Rawlings. And then there are the philosophers: Thomas Paine, John Locke, John Stuart Mill, John Keynes, Edmund Burke and Adam Smith. The international impact of British musicians and composers such as the Rolling Stones, the Kinks, the Beatles, Queen, Pink Floyd, Holst and Elgar, cannot be denied. The list is endless.

The story of British creative success is a great topic for historians. But it also holds great lessons for the future. It is a lesson in how to construct an innovative cultural and social system that sustains high quality output over centuries. Britain is indeed the global "mother of invention," and we would do well to better understand and preserve those forces that have allowed it to contribute so much to humankind.

About the author

Keith grew up in a relatively bookless home. But this paucity of reading matter acted as a catalyst to discover, encouraging him to explore museums, libraries and ancient sites in his locality to quench a thirst for all things historic. An entrepreneur by nature, he started his first business at the age of 11, buying jug and basin sets and wind-up gramophones from Jumble sales and after refurbishment, sold them on to his developed network of antique and second-hand dealers in Dover and more latterly Oxford where he later resided with his family.

His main subject interests were in music, English, history, and science. When he left school, he pursued his dream of becoming an electrical engineer. Later, mid-career, he studied Business Administration gaining a master's degree and is currently planning and researching towards a Ph.D. Keith has held many senior positions in start-up ventures with large corporations, including mobile telecom giant Orange and a Pan European Data Centre company. He has four grown-up daughters, lives in the Cotswolds and is currently creating and collaborating on ground-breaking developments within the green-tech and mixed energy sectors. He is an ardent amateur modern industrial historian, business guru and a Fellow of the Chartered Management Institute (FCMI) and a Member of the Institution of Engineering and Technology (MIET). In his spare time, Keith enjoys visiting anywhere by the ocean, loves to cook and submits to the paradigm that continuous learning is the key to longevity. Keith's extensive track record of success and dedication to his profession makes him a reliable source of inspiration for the younger generation.

Chronological index by year

Year	Page	
3000	BC. Circa. English language.	29
3000	BC. Stonehenge	201
2500	BC. Avebury Stone Circles	196
10	AD. Dover Castle	241
10	AD. Halloween	166
70	AD. Roman Baths – in Bath.	200
100	AD. Circa. Haggis.	80
461	AD. Saint Patrick	244
500	AD. King Arthur.	241
1000	AD. Circa. Loch Ness and Nessie.	241
1000	AD. Circa. Silver ball hurling. Cornwall	244
1066	Tower of London	201
1068	Warwick castle	202
1070	Windsor Castle	202
1072	The Houses of Parliament. Westminster Abbey	197
1087	Tewkesbury Abbey	201
1090	Bath Abbey.	197
1096	Oxford University	202
1130	Edinburgh Castle	198
1150	St Michaels Mount – Cornwall	200
1166	Jury Trial	66
1170	America (USA) discovery	41
1200	Circa. Guillotine. Town of Halifax	55
1200	Leeds Castle	199
1209	Cambridge University	202
1215	Magna Carta. World's first bill of rights	66
1215	Habeas Corpus	65
1236	Parliamentary democracy	68
1250	Magnifying glass	181
1283	Caernarfon Castle	198
1295	Beaumaris Castle	197
1299	Bowls – the sport.	188
1300	circa. Full English breakfast.	80

1331	Patent. first documented granting of	17
1377	Robin Hood. Legend	243
1386	Clock. World's oldest working clock. Salisbury	20
1400	circa. Golf	190
1400	circa. Cheese rolling.	241
1400	circa. Rounders.	192
1400	circa. Quakers. George Fox	69
1417	Street Lighting. Sir Henry Barton	228
1475	English book – first printed	131
1480	circa. Sunday Roast.	81
1483	Oil – the barrel measurement. James Young	108
1494	Hampton Court Palace	198
1500	circa. Utilitarianism. Jeremy Bentham	70
1534	Cambridge University Press	129
1509	Cornish pasty	79
1557	Equal sign. Robert Recorde	108
1564	William Shakespeare. Playwright and dramatist.	134
1578	Submarine – first navigable. William Bourne	118
1583	Commonwealth of nations. Development.	42
1589	Knitting machine. Reverend William Lee	73
1596	Toilet – flushing. John Harrington	91
1600	circa. Anglicanism. Henry VIII	64
1600	Magnetic earth. William Gilbert	181
1603	Railway tracks – wooden. Huntingdon Beaumont	214
1605	Bacons Cipher. Sir Francis Bacon	31
1605	Bonfire night. November 5th	240
1614	Logarithms. John Napier	108
1617	One Way Street. London	225
1622	Slide rule. William Oughtred	109
1625	Taxi – Hackney cab	228
1627	Water desalination – first experiments	144
1628	Blood circulation – discovery of. William Harvey	165
1630	Telescopic sight. William Gascoigne	59
1632	Liberalism - Classic. John Locke	67
1635	Champagne bottle design. Robert Mansell	78
1638	Micrometer. William Gascoigne	56
1650	Coffee houses. Oxford.	79

1655 Spectrum – heterogeneity of light. Sir Isaac Newton 184
1660 Balance spring. Robert Hook 20
1660 Daniel Defoe. Novelist 130
1661 Postmarks. Henry Bishop 23
1662 Sparkling wine (Champagne). Christopher Merrett 77
1663 Red blood cells. Robert Hooke 149
1664 Binary stars. Robert Hooke 145
1664 10 Downing Street. Built by Sir Christopher Wren 196
1665 Compound microscope. Robert Hooke 156
1665 Gravity – the law of. Sir Isaac Newton 176
1668 Metric system. John Wilkins 108
1668 Telescope – reflecting. Sir Isaac Newton 184
1669 Calculus. Isaac Newton 107
1674 Lead crystal glass. George Ravenscroft 102
1676 Universal Joint. Robert Hooke 61
1680 Postage stamps. William Dockwra 24
1680 Sign Language. George Dalgarno 24
1684 Fingerprinting. Nehemiah Grew 30
1686 Jet propulsion – theory. Sir Isaac Newton 178
1687 Motion – the law of. Sir Isaac Newton 182
1689 Chocolate – Drinking chocolate. Hans Sloane 78
1689 Kensington Palace. 199
1691 Diving Bell. Edmund Halley 230
1698 Steam Engine. Thomas Savery 58
1700 circa. Mass Production 17
1700 circa. Industrial revolution starts. 94
1700 Circa. Manufacturing revolution starts 64
1700 circa. Methodism. John and Charles Wesley 67
1700 circa. Welsh Rarebit 82
1705 Blenheim Palace 135
1701 Seed drill. Jethro Tull 14
1704 Halley's comet. Edmond Halley 146
1705 Steam Engine. Atmospheric 59
1705 Blenheim Palace 197
1709 Blast furnace. Abraham Darby 99
1710 St Pauls Cathedral 200
1711 Tuning Fork. John Shore 122

1711 Car – Powered. Christopher Holtum 220
1714 Typewriter. Henry Mill 136
1718 Machine gun. James Puckle 114
1719 Three choir's festival 122
1720 English Horn. 120
1720 circa. Maritime Clock. John Harrison 140
1723 Water sprinkler. Ambrose Godfrey 48
1725 Stereotyping. William Ged 136
1727 Bridge – Railway. Ralph Wood 51
1727 Surgical forceps. Stephen Hales 170
1728 Speed of light – discovery. James Bradley 184
1730 Spectacles – modern side arm. Edward Scarlett 143
1733 Pram – perambulator or baby carriage. William Kent 89
1733 Blood pressure measurement. Stephen Hales 165
1747 Controlled clinical trials. James Lind 157
1748 Fridge – principal of heat exchange. William Cullen 86
1750 Horse racing – thoroughbred. 191
1752 Sunglasses. James Ayscough 143
1755 Dictionary. Dr Samuel Johnson 131
1755 Baseball. William bray 187
1775 Jane Austin. Novelist 133
1761 Buckingham Palace 198
1761 Chronometer – marine. John Harrison 230
1762 Sandwich. John Montague - The Earl of Sandwich 54
1763 Aspirin – discovery of active ingredient. Edward Stone 164
1764 Spinning Jenny. James Hargreaves 75
1767 Railway tracks – Cast Iron. 214
1767 Modern Sociology. Adam Ferguson 68
1767 Jigsaw. John Spilsbury 237
1768 Spinning frame. Richard Arkwright 76
1769 New Zealand. Discovery and settlement 46
1770 Toothbrush. William Addis 92
1770 Australia formation. Captain James Cook 42
1770 Caterpillar Tracks - Patent. Richard Edgeworth 52
1770 Eraser. Edward Naime 86
1772 Nitrous oxide – discovery of. Joseph Priestly 153
1772 Figure skating 190

1772 Fizzy drinks. Joseph Priestly 80
1774 Lifeboat. Lionel Lukin 48
1774 Concrete – quick drying. John Smeaton 100
1774 Oxygen – discovery of. Joseph Priestly 153
1774 Cricket – first rules. Worlds second most popular sport. 188
1775 Toilet – S-trap. Alexander Cummings 91
1776 Lifeboat station. 48
1776 Threshing machine. Andrew Meikle 15
1776 Modern economics. Adam Smith 68
1776 Hydrogen – discovery of. Henry Cavendish 152
1778 Tumbler lock. Robert Barron 60
1779 Spinning mule. Samuel Crompton 76
1779 Bridge – Iron. Abraham Darby 51
1780 Steel Pen. Samuel Harrison 90
1780 Circus – World's first. Philip Astley. 236
1781 Uranus – discovery of. William Herschel 147
1782 Steam Engine – modern condensing 59
1782 Sandwich. Earl of Sandwich 80
1783 Clouds – Categorisation of. Luke Howard 95
1784 Tumbler lock – high security. Joseph Bramah 61
1785 Loom – powered. Edmund Cartwright 74
1788 Flyball Governor. James Watt 54
1788 George Byron. Novelist 132
1790 Shoelaces. Harvey Kennedy 89
1790 Sewing machine. Thomas Saint 75
1791 Gas Turbine. John Barber 54
1792 Illuminating gas. William Murdoch 177
1794 Street Lighting – gas. William Murdoch 228
1794 Ball Bearing. Phillip Vaughan 49
1794 Colour blindness – discovery of. John Dalton 156
1795 Corkscrew. Samuel Henshall 85
1795 Hydraulic press. Joseph Bramah 102
1796 Vaccination. Edward Jenner 172
1797 Banknotes. Bank of England 85
1799 Anaesthetics – nitrous oxide. Sir Humphrey Davy 162
1800 circa. Seat Belt. Sir George Cayley 58
1800 circa. Polo – the game. 192

1800 circa. Billiards. 188
1800 circa. Wellington Boots. Duke of Wellington 63
1800 circa. Pearly kings and queens. 243
1800 Light bulb – incandescent. Sir Humphrey Davy 179
1800 circa. Hockey 190
1801 Steam engine. Richard Trevithick 215
1801 Daltons Law – of partial pressures. John Dalton 174
1802 Clouds – categorisation of. Luke Howard 140
1804 Rocket – solid fuel. William Congreve 116
1804 Steam Locomotive. Richard Trevithick 215
1805 Atomic Theory. John Dalton 139
1807 Chemical electrolysis. Sir Humphrey Davy 152
1807 Street Lighting. Samuel Clegg 228
1807 Potassium – first isolation of. Sir Humphrey Davy 153
1807 Sodium – First isolation of. Sir Humphrey Davy 154
1808 Tension spoked wheel. Sir George Cayley 60
1808 Aluminium – the discovery of. Humphrey Davy 98
1808 Boron – isolation of 152
1810 Tin Can. Peter Durand 81
1814 Charles Dickens. Novelist 130
1815 Safety Lamp. Sir Humphrey Davey 57
1815 DH Lawrence. Novelist 130
1816 Fire Extinguisher. Modern portable 47
1816 Charlotte Brontë. Novelist 130
1816 Kaleidoscope. Sir David Brewster 178
1817 Parkinson's disease – discovery of. James Parkinson 161
1818 Emily Jane Brontë. Novelist 131
1818 Blood transfusions – human to human. James Blundell 165
1818 Ice hockey. 191
1819 George Elliot. Novelist 132
1819 Hay Fever – discovery of. John Bostock 159
1820 Computer. First mechanical computer 35
1820 circa. Gin & Tonic 80
1820 Hygrometer. John Frederic Daniell 102
1820 Macadam. John Loudon McAdam 103
1820 Anne Brontë. Novelist 129
1820 Squash. Harrow – public school. 194

1821 Electric transformer. Michael Faraday 174
1823 Lift or Elevator. T Horner and D Burton 56
1823 Mackintosh – waterproof coat. Charles Mackintosh 74
1823 Lancet – first medical journal 133
1823 Electro Magnet. William Sturgeon 175
1823 Rugby Union 193
1824 Fire Brigade. James Braidwood. 47
1824 Cement. Joseph Aspdin 99
1825 Benzene – first isolation of. Michael Faraday 151
1825 Railway – Passenger. George Stevenson 213
1826 Stove – gas. James Clegg 90
1827 Lawnmower. Edwin Beard Budding 87
1827 Friction match. John Walker 86
1828 Mechanical reaping machine. Patrick Bell 14
1828 Thermosiphon. Thomas Fowler 60
1829 Police. Bow Street Horse Patrol. 47
1829 Concertina. Sir Charles Wheatstone 120
1830 Adjustable spanner or wrench. Edward Budding 49
1830 Thermostat. Andrew Ure 60
1830 Railway – first regular steam passenger service 213
1830 Elastic fabric. Thomas Hancock 73
1830 Railway – first inter-city. Liverpool to Manchester 213
1831 Bus Service – inter-city 219
1831 Safety fuse – explosives. William Bickford 104
1831 Electric Generator. Michael Faraday 174
1831 Electric motor. Michael Faraday 174
1832 Waterbed. Neil Arnott 93
1832 Thomas Hardy. Novelist 136
1832 Dental chair – fully adjustable. James Snell 158
1833 Sandpaper. John Oakey & Sons Ltd 104
1834 Railway – Season ticket 214
1834 Venn diagram. John Venn 110
1834 Mail order. Pryce Jones 16
1835 Photography – practical. William Henry Fox Talbot 127
1836 Arsenic test. James Marsh 28
1836 Screw Propeller. Sir Francis P Smith 57
1836 Faraday Cage. Michael Faraday 175

1837 Telegraph. First electric telegraph 24
1837 Shorthand. Sir Isaac Pitman 135
1837 Railway tickets. Thomas Edmondson 214
1838 Typhoid – discovery and prevention. William Budd 170
1838 Internal Combustion Engine. William Barnett 55
1838 Stereoscope 3D. Sir Charles Wheatstone 127
1838 Ship – Steam. Isambard Kingdom Brunel 232
1838 Ship – propeller driven. The Archimedes 232
1839 Steam Hammer. James Nasmyth 105
1839 Bicycle - Safety with pedals. Kirkpatrick MacMillan 218
1839 Timetable – national 136
1840 Electro plating 100
1840 Water and sewage systems. William Lindley 106
1840 Boolean Algebra – Digital logic design. George Boole 107
1841 Railway – first signals – semaphore 214
1841 Screw threads - Standardisation. Joseph Whitworth 57
1841 Clock. First electric. Alexander Bain 21
1841 Hypnosis. James Braid 167
1842 Fuel Cell – William R Grove 175
1843 Steel hulled ship – screw propeller 233
1843 Greeting cards. John Callcott Horsley 23
1843 Computer programme. First programmer. Ada Lovelace 36
1843 Conservation of energy – law of. James Joules 173
1843 Fax Machine - patent. Alexander Bain 38
1843 Joule – Thermodynamics. James Prescott 178
1845 Rubber band – Stephen Perry 57
1845 Tyre – pneumatic. Robert W Thompson 61
1845 Diamagnetism. Michael Faraday 174
1846 Free trade theory. Dr Andrew Ure 65
1847 Hydraulic Crane. George Armstrong 55
1847 Chocolate bar – first. Joseph Fry 78
1847 Neptune – discovery of. John Couch Adams 146
1847 Anaesthetics – Chloroform. Sir James Young Simpson 163
1848 Bridge – box girder. Robert Stephenson 50
1848 Zero – devised scale to show zero. William Thompson 185
1848 Chain Stores 16
1848 Aircraft – First powered flight. John Stringfellow 204

1849 Safety Pin. Charles Rowley 89
1849 Aircraft - First heavier than air flight. Sir George Cayley 205
1850 Transatlantic Cable. John Watkins Brett. 26
1850 Mercerized cotton. John Mercer 75
1850 Bessemer converter – mass production of steel 99
1850 circa. Snooker 194
1850 Robert Louis Stevenson. Novelist 134
1851 Oil refinery. James Young 103
1851 Oil – first oil tycoon. James Young 103
1851 Fax Machine – developed working model 38
1851 Toilet – public. John Neville Maskelyne 91
1852 Thesaurus. Peter Mark Roget 136
1853 Aquarium. Phillip Gosse 49
1853 Aircraft – First manned controlled flight 205
1853 Hypodermic syringe. Alexander Wood 160
1854 Pie Chart. Florence Nightingale 109
1854 Cipher (Playfair). Charles Wheatstone 31
1854 Concrete – reinforced. William Boutland Wilkinson 100
1854 Cholera – discovery that it was water bound 166
1855 Plastic. Alexander Parkes 141
1855 Package holiday. Land travel 238
1856 Radio – discovery. David Edward Hughes 183
1856 Dye – synthetic. William H Perkin 72
1856 Dinosaur – nomenclature. Richard Owen 148
1857 Railway tracks – Steel. Matthew Kirtley 214
1857 Postcodes or Zip Codes. London. 23
1857 Edward Elgar – composer 120
1857 Female Doctor - first. Dr Elizabeth Blackwell 167
1858 Epilepsy – pioneer in treatment of 159
1858 Gray's Anatomy. Henry Gray 167
1859 Evolution – theory of. Charles Darwin 148
1859 Arthur Conan Doyle. Novelist 129
1860 Linoleum. Frederick Walton 102
1860 JM Barry. Novelist 133
1860 Fish and chips 79
1861 Thallium – discovery of. William Crookes 154
1862 Balmoral Castle. 196

1863 Broadmoor hospital 28
1863 Football – Football Association (FA) 189
1863 Railway – Underground. 216
1864 Microwave – discovery. James Clerk Maxwell 182
1865 Rudyard Kipling. Novelist 133
1865 Sewage Systems. Joseph Bazalgette 96
1865 Periodic table – John Alexander Reina Newlands 153
1865 Antiseptic surgery. Joseph Lister 163
1865 Boxing – Queensbury – current rules. John Chambers 188
1866 Torpedo. Robert Whitehead 118
1866 Beatrix Potter. Novelist 129
1866 HG Wells. Novelist 132
1868 Umbrella. Samuel Fox 92
1868 Ailerons. Matthew Watt Boulton. 204
1868 Standard deviation – probability 109
1868 Helium – First observed. Sir Joseph Lockyer 152
1868 Traffic Lights – Gas. John Peake-Knight 229
1870 Bicycle – Modern 218
1870 Bathtub – cast iron enamel bath. David Buick 85
1870 Lorry or Truck. John Yule 224
1870 Plimsol line. Samuel Plimsol 232
1871 Wind tunnel. Francis Herbert Wenham 211
1872 Holiday pay. George Livesy 66
1872 Football. The FA Cup. 189
1872 Ralph Vaughan Williams 123
1873 Light – nature of electromagnetic theory 183
1873 Badminton 186
1873 Bicycle – Chain drive. Henry John Lawson 217
1874 Tennis. Major Harry Gem 195
1874 Electric cooking Utensil. St. George Lane-Fox 86
1874 Gustav Holst – composer 121
1876 Telephone. Alexander Graham Bell 25
1877 Bridge – concrete. Philip Brannan 50
1878 Microphone. David E Hughes 23
1879 Cathode Ray Tube. Sir William Crookes 139
1879 Blackpool illuminations. 235
1880 Hydroelectricity. William Armstrong 177

1880 Light bulb. Joseph Swan 179
1880 circa. Table Tennis or Ping-pong 194
1982 Virginia Woolf. Novelist 137
1883 Whistles. Joseph Hudson 27
1883 Fibre – man-made. Joseph Swann 73
1884 Light switch – Quick break technology 179
1884 Multiple-wheel steam turbine 36
1884 Motorcycle – first petrol powered. Edward Butler 225
1884 Greenwich meridian. GMT 231
1884 Steam Turbine. Sir Charles Parsons 59
1884 Car – Electric. Thomas Parker 220
1885 Salvation Army. William Booth 69
1885 Caravan. Dr Gordon Stables 222
1886 Heart lung and liver transplant 108
1886 Tower bridge – London 201
1886 Chocolate bar – Kit-Kat is world's most popular 78
1887 Pneumatic tyres – air inflated. John Boyd Dunlop 219
1887 Clinical thermometer. Sir Thomas Allbutt 158
1887 Theatre organ. Robert Hope-Jones 122
1887 Bicycle tyre – Pneumatic 150
1888 Spark plug. Sir Oliver Lodge 58
1888 Solar Power – First solar cell patent. Edward Weston 183
1888 Tiddlywinks. Joseph Fincher 195
1889 Cinematography – creator. William Friese Greene 125
1890 Diesel Engine – pre-Diesel patent 54
1890 Agatha Christie. Novelist 128
1891 Kettle – electric. Compton & Co Ltd. 87
1891 Synthetic rubber – Isoprene 144
1891 Basketball. James Naismith. 187
1892 Netball 191
1892 Rayon – Viscose 75
1892 Vacuum flask. James Dewar 93
1892 Newspaper – first printed in Braille 134
1893 Lawnmower – Steam powered. James Sumner 87
1893 Railway – signals – electric. 214
1893 Toaster – electric 90
1894 Argon – discovery and proof 151

1895 Rayon – Acetate 75
1895 Rugby League. 193
1896 Seismograph. John Milne 142
1896 Typhoid – first vaccine. Almroth Edward Wright 171
1896 Darts – board and rules. Brian Gamlin 189
1897 Electron – discovery of. JJ Thomson 175
1897 Ship – Steam Turbine. Royal Navy 117
1897 Enid Blyton. Novelist 131
1897 Malaria – discovery of cause. Ronald Ross 169
1898 Lorry – Articulated. Thornycroft Co Ltd. 224
1898 Meccano. Frank Hornby 88
1898 Polyethylene. Eric Fawcett and Reginald Gibson 141
1899 Plasticine. William Herbert 141
1899 Hydrofoil. John Isaac Thorneycroft 232
1899 Aircraft – First manned powered controlled flight 205
1899 Silicone. Frederick Kipping 143
1899 Mousetrap. James Henry Atkinson 89
1900 Cartoon. J Stuart Blackton 125
1901 Tractor. Daniel Albone 15
1901 Women's Liberation. Emily Davison 71
1901 Vacuum cleaner. Herbert Cecil Booth 92
1901 Tarmac. Edgar Purnell Hooley 105
1902 Tea Maker - automatic. Albert E Richardson 90
1902 Disc Brakes. Frederick William Lanchester 222
1903 Caterpillar tracks – first commercial use 52
1903 George Orwell. Novelist 132
1903 Hormones – Discovery of. William Bayliss 149
1904 Wendy – the name. JM Barry 137
1904 Spare Wheel. Thomas and Walter Davies 227
1904 Diode Rectifier Tube – Radio Valve 38
1905 Rolex Watch Company 17
1906 Kinemacolour – colour movie process 126
1906 ARG Electrically powered analogue computer 34
1906 Dreadnought – first turbine powered battleship 113
1906 Rolls Royce. 226
1906 Holiday Camp. Dodd's 237
1907 LED – theory of. HJ Round 179

1907 Scouting. Robert Baden-Powell 164
1907 Racing car circuit. Brooklands. Hugh Locke-King 225
1908 Television. John Logie Baird. 25
1908 London. First electric advertising 200
1908 Ian Fleming. Novelist 132
1909 Bus – Double decker. 219
1909 Mini Roundabout. Frank Blackmore 224
1910 Wall plug. John J Rawlings 62
1911 Atom - discovery of. Ernest Rutherford 138
1912 SONAR. Lewis Fry-Richardson 117
1912 Scooter – child's. C E Richardson & Co Ltd 227
1912 Bragg's Law. William Lawrence Bragg 173
1912 Isotopes – first concept of. Frederick Soddy 177
1913 Stainless Steel. Henry Brearly 105
1913 Shaver – electric. GP Appleyard 89
1913 Aston Martin. Robert Banford and Lionel Martin 217
1913 Crosswords. Arthur Wynne. 236
1914 Tank – military. Ernest Swinton 117
1914 Theatre Organ. Robert Hope-Jones 81
1914 Fighter Aircraft. Vickers company 113
1914 Depth charge. Herbert Taylor 112
1914 Acetylcholine – discovery of. Henry Hallet 162
1916 Roald Dahl. Novelist 134
1917 Aircraft – landing on a moving ship. EH Dunning 206
1918 Air Force. World's first independent air force. RAF 209
1918 RAM Random Access Memory 40
1918 Gas Mask. Edward Harrison 112
1918 Aircraft carriers – HMS Argus 111
1919 Mass Spectrograph. Sir William Aston 181
1919 Package holiday. Air travel 238
1919 First International daily air passenger service 210
1920 Folding carton. Charles Henry Foyle 66
1921 Blood transfusion service. Dr Percy Oliver 165
1921 Vehicle Road Tax. 229
1922 BBC 234
1922 Kingsley Amis. Novelist 132
1922 Jaguar – luxury sports cars. 223

1923 Wristwatch. First automatic. John Harwood 21
1925 Television. First television. John Logie Baird 25
1928 Television. Transatlantic transmission 25
1928 Television. First colour TV. John Logie Baird. 25
1928 Video Recorder. John Logie Baird. 26
1928 Penicillin - discovery. Sir Alexander Fleming 169
1930 Jet Engine. Sir Frank Whittle 55
1931 Stereo Sound. Alan Dower Blumlein 24
1931 Geological Timescale. Arthur Holmes 140
1932 Splitting the atom 182
1932 Neutron – discovery of. James Chadwick 124
1934 Cats eyes – road. Percy Shaw 222
1935 RADAR – radio locator. Sir Robert Watson-Watt 114
1935 Hearing aid. Edwin Stevens 160
1935 Aerial refuelling. Sir Alan Cobham 203
1936 Keynesianism. John Maynard Keynes 67
1937 Emergency calls 999 22
1938 Train – Steam. World's fastest. The Mallard 216
1938 Game show for TV. Spelling Bee 237
1939 RADAR – Airborne. Alan Blumein 115
1939 Bletchley Park – code breakers 32
1940 Ejector seat. Sir James Martin 209
1941 Special forces – SAS and SBS 116
1941 Polyester. John Whinfield & James Dixon 73
1942 Head up Display. Cintel 209
1943 Computer – Electronic programmable 37
1943 Bouncing bomb. Dr Barnes Wallis 111
1945 Penicillin – mass production 169
1945 Satellites – geostationary. Sir Arthur C Clarke 142
1945 JCB – earth moving equipment. Joseph Cyril Bamford 223
1946 Formula 1 – Motor racing 189
1947 Land Rover. 224
1948 Computer – first stored programme 36
1948 NHS – National Health Service. Aneurin Bevan 161
1948 Paralympic Games 192
1948 Aircraft – Worlds first turboprop airliner 206
1949 Aircraft – first jet airliner. De Havilland Comet 210

1949	Biometrics. Iris recognition. James Doggart	35
1949	Soft whipped ice cream	81
1949	Jet bomber. English electric company	113
1949	Cheltenham Literature festival	130
1949	Cataracts – artificial lens implant. Harold Ridley	155
1950	Smoking and lung cancer link	169
1951	Anaesthetics – Halothane. Charles Suckling	163
1952	Integrated Circuit. GWA Dummer	39
1952	Trackball. Kenyon Taylor	40
1952	Float Glass. Alistair Pilkington	101
1952	Mousetrap – world's longest running play	237
1953	DNA – discovery	166
1954	Fibre Optics. John Tyndall	22
1954	Aircraft – Worlds first VTOL. Short SC1	206
1955	Hovercraft – amphibious. Christopher Cockerell	231
1955	Clock. First atomic clock. Louis Essen	21
1955	Kettle – Automatic electric. Russell Hobbs Co Ltd	87
1956	Nuclear power – first generating station	183
1956	Clean air act.	65
1956	Ultrasound imaging. Ian Donald	171
1958	Sailboard and sail boarding. Peter Chilvers	193
1960	circa. Stun Grenades. SAS	117
1960	The Beatles	119
1960	Coronation Street – longest running soap opera.	236
1961	Hip replacements. Sir John Charnley	160
1961	Amnesty International. Peter Benenson	64
1961	Ibuprofen. Stewart Adams – Boots PLC	168
1962	Razor – long-life blades. Wilkinson sword	56
1962	Cloning – theory of. John B Gurdon	156
1963	Lava Lamp. Edward Craven Walker	88
1963	Carbon Fibre	99
1964	Beta blockers – propranolol. James Black	164
1965	Aircraft – VTOL first commercial	207
1965	Packet switching. Donald Davies	39
1965	Baby Buggies – Strollers. Owen MacLaren	83
1965	Yesterday – The Beatles. Most covered song	123
1965	JK Rowling. Novelist	133

1965 Portable defibrillator. Dr James Pantridge 161
1966 Mini skirt. Mary Quant 17
1966 Safari Park – first drive through safari park 238
1966 Car - First permanent 4-wheel drive 221
1966 Aspirin – effectiveness. John Vane 111
1966 ATM (cash machine) PIN. James Goodfellow 33
1966 Car – Standard anti-lock braking. Jenson interceptor 221
1967 ATM Cash machine. John Shepherd Barron 33
1967 Pulsars – Discovery. Jocelyn Bell 146
1968 Workbench – portable. Ron Hickman 63
1968 Contraceptive pill. Herchel Smith 157
1969 Aircraft – Commercial supersonic airliner. Concorde 207
1970 LCD – liquid Crystal Display. George Gray 180
1970 Range Rover. 226
1970 Glastonbury – largest music festival 120
1970 Bar Codes (Industrial). Plessey 34
1971 CAT or CAT Scanner. Godfrey N Hounsfield 155
1972 Calculator – pocket. Sir Clive Sinclair 35
1973 RSA Cipher. Clifford Cocks 32
1973 MRI – Magnetic Resonance Imaging 168
1974 Ball barrow. James Dyson 84
1974 Black holes. Stephen Hawking 145
1975 Tilting Train. British Rail Derby 215
1975 Solar powered boat. Alan T Freeman 232
1976 Train – Diesel. World's fastest 215
1976 Lithium-ion battery. John Goodenough et al. 180
1977 Test tube baby 149
1978 IVF. Patrick Steptoe and Robert Edwards 168
1978 ARM Universal mobile device processor 34
1979 Computer - Laptop. Bill Moggridge 36
1979 iPod origins. Kane Kramer 39
1980 MRI – Machine 168
1981 Computer - Portable. Adam Osbourne 36
1982 Diabetes – first synthetic insulin 158
1982 Aspirin – effectiveness 164
1983 Television. Pocket TV. Sir Clive Sinclair 26
1983 Bagless Vacuum cleaner. James Dyson 84

1984	DNA fingerprinting. Alan Jeffreys	29
1984	Computer - handheld. Psion	37
1984	MAGLEV- first commercial MAGLEV Train	212
1985	Live Aid	121
1986	Heart lung and liver transplant	160
1986	Monoclonal anti-bodies	149
1987	Robert Melias. First person convicted of crime	30
1989	HTTP-URL's-HTML. Sir Tim Berners-Lee	38
1989	WWW World Wide Web. Sir Tim Berners-Lee	40
1990	Web Browser. Sir Tim Berners-Lee	40
1990	Anywayup cup. Mandy Haberman	83
1991	Viagra. Andrew Bell, David Brown	154
1991	Wind-up radio. Trevor Bayliss	184
1992	SMS. First SMS to be sent phone to phone	24
1994	Iris recognition. John Daugman	35
1994	Channel Tunnel – Anglo French.	52
1995	DNA Database. UK Home office	29
1996	Cloning – first successful. Dr Ian Wilmutt	156
1997	Hard Drives. Professor Stuart Parkin	38
1997	Car – World's fastest. Thrust SSC.	220
1997	Teletubbies. Anne Wood and Andrew Devonport	239
1998	Bus-hybrid. BAE Systems	219
1998	Who wants to be a millionaire? David Briggs et al.	239
2000	Wealth creation.	18
2000	Banknotes. The world's biggest manufacturer of	31
2001	iPod. Sir Jonathon Ive	39
2001	Bridge – world's tallest (GB design)	51
2001	Bridge – Tilting	52
2004	Graphene. Professor Andre Geim	176
2005	Liquid Armour. BAE Systems	114
2005	Bicycle – first lightweight Lithium-ion powered	218
2006	Batteries – structural. BAE Systems	98
2006	Car – World's fastest Diesel. JCB Diesel Max	220
2007	Britain's got talent. Simon Cowell	236
2008	Face recognition. BAE Systems	29
2008	Laser timing. BAE Systems	113
2008	Stem Cells – first practical use. Martin Burchall	149

2008	Steri-Spray – sterilises water. Ian Helmore	170
2009	Aircraft – First autonomous aircraft. Mantis	208
2014	Vantablack® - worlds blackest material	62
2014	Brain 'locator GPS'. Dr Percy Oliver	166
2019	Car – World's fastest future car. Bloodhound SSC	221
2020	SABRE HOTOL – Hypersonic jet/rocket	210

Printed in Great Britain
by Amazon

69238851R00154